METAL TOXICITY IN MAMMALS•1

Physiologic and Chemical Basis for Metal Toxicity

METAL TOXICITY IN MAMMALS

Volume 1 • *Physiologic and Chemical Basis for Metal Toxicity*

Volume 2 • *Chemical Toxicity of Metals and Metalloids*

METAL TOXICITY IN MAMMALS • 1

Physiologic and Chemical Basis for Metal Toxicity

T. D. LUCKEY
AND
B. VENUGOPAL

Department of Biochemistry
University of Missouri, Columbia

PLENUM PRESS · NEW YORK AND LONDON

Library of Congress Cataloging in Publication Data

Luckey, Thomas D
   Metal toxicity in mammals.

   Bibliography: v. 1, p.
   Includes index.
   CONTENTS: v. 1. Physiologic and chemical basis for metal toxicity.
   1. Metals—Toxicology. 2. Mammals—Diseases. I. Venugopal, B., joint author. II. Title.
RA1231.M52L82   599'.02'4   76-44859
ISBN 0-306-37176-2 (v. 1)

First Printing—January 1977
Second Printing—March 1979

© 1977 Plenum Press, New York
A Division of Plenum Publishing Corporation
227 West 17th Street, New York, N.Y. 10011

All rights reserved

No part of this book may be reproduced, stored in a retrieval system, or transmitted, in any form or by any means, electronic, mechanical, photocopying, microfilming, recording, or otherwise, without written permission from the Publisher

Printed in the United States of America

PREFACE

Our purpose is to provide understanding for appropriate use of metals in a technical society. Knowledge of metal toxicity is needed for the prevention, prediction, diagnosis, and therapy of adverse reactions from excess metals in mammals. *Metal Toxicity in Mammals* is presented in two volumes. Volume 1, *Physiologic and Chemical Basis for Metal Toxicity,* provides the basis for understanding the toxic actions of metals recorded in Volume 2, *Chemical Toxicity of Metals and Metalloids.* The details and bases for many concepts summarized in Volume I are given, with appropriate references, in Volume 2. Thus, references for specific items in Volume 2 are not generally given in Volume 1.

The authors reviewed the known toxicity of several heavy metals in anticipation of their use as multinutrient markers for NASA. As more and more metals were considered, the need for a complete review became obvious. This treatise supplants onerous searches of metal-toxicity literature up to 1975 and reviews the toxicity of all the metals of the periodic table on the basis of available relevant data. Books on pharmacological, nutritional, medical, veterinary, or industrial toxicity contain information about selected metals. More complete data about metals of public concern, such as mercury, lead, and cadmium, may be found in numerous books and reviews. The reader should refer to general texts and basic reference works, when specific references are not given, for general information.

The authors were mindful of those physical scientists who have had little formal education in biology and of those biologists who have had little opportunity to use their textbook knowledge of metal chemistry. We hope all will be excited by the challenge of greater understanding when a working knowledge of both disciplines has been acquired. This is the rationale for using chemical names in Volume 1 and symbols in Volume 2; exceptions include ions, organic complexes, and tables. Both volumes contain the atomic table and an alphabetical listing of the elements. A glossary is appended (Dorland, 1974). The summary of atomic structures near the beginning of each chapter in Volume 2 helps to show how the anatomy of the atom directs its chemical reactions to give the biologic activities reported or predicted for each metal. The rudimentary base of physiology and chemistry provided in Volume 1 helps the reader achieve this synthesis. This, plus knowledge of biochemical reactions for a few metals, provides a general understanding of possible mechanisms of toxicity at the molecular level that can be envisioned when there is little or no specific information.

A summary of metal toxicity by groups in the periodic table is given in the last chapter of Volume 1. Toxicity for each metal by all modes is given in the appropriate chapter in Volume 2, and comparative toxicity of metals by groups, periods, modes of administration, and species, is given in the summary chapter of Volume 2.

The authors are indebted to proofreader Pauline Luckey and to general readers Dr. Doug Frost, Nutrition Consultant, Schenectady, New York; Dr. Gary Van Gelder, Veterinary Toxicologist at the University of Missouri-Columbia (UMC); Dr. Howard Hopps, Pathologist at UMC; Dr. Dave Hutcheson, Associate Professor of Animal Nutrition and Statistician. University of Missouri Sinclair Comparative Medicine Research Farm; and especially to Dr. Mike Kay, Analytical Chemist at UMC Nuclear Reactor. Each has provided helpful criticism of both style and substance, but the authors of course claim responsibility for any shortcomings which remain.

CONTENTS

Chapter 1
INTRODUCTION TO HEAVY METAL TOXICITY IN MAMMALS   1

Scope and Limitations 3
Essential Metals 5
Stimulatory Metals 9
Pharmacologic and Therapeutic Use of Metals 10
Toxicology 13
Safety 20
Analytic Procedures 26
    Sample Preparation 27
    Analytic Techniques 28
        Atomic Absorption 28
        Emission Spectroscopy 29
        X-Ray Fluorescence 30
        Electron Microprobe 31
        Gas–Liquid Chromatography 32
        Spark-Source Mass Spectrometry 34
        Anodic Stripping Voltametry 34
        Neutron Activation 34
    Summary of Techniques 36

Chapter 2
MODES OF INTAKE AND ABSORPTION 39

Enteral Administration 43
    Oral Intake and Absorption 43
    Gastrointestinal Absorption and Persorption 43
        Introduction 43
        The Villus 46
        Absorption 52

Phagocytosis	57
Persorption	57
Gavage	58
Aboral Administration	58
Inhalation	59
The Respiratory Tract	59
Inhaled Pollutants	64
Particulate Deposition	68
Particle Uptake and Removal	72
Absorption	74
Physiologic Effect of Particle Size	75
Translation of Inhalation Data	76
Dermal Absorption and Administration	80
Other Parenteral Routes	86
Intravenous Administration	87
Intraperitoneal Administration	88
Intramuscular Administration	88
Vaginal Administration	89
Other Routes	89
Summary of Modes	91

Chapter 3

DETOXICATION, EXCRETION, AND PHYSIOLOGIC HOMEOSTASIS 93

Detoxication	94
Excretion	97
Homeostasis	99

Chapter 4

TOXICOLOGIC SIGNIFICANCE OF THE PHYSICO-CHEMICAL PROPERTIES OF METALS 103

Physicochemical Properties of Biologic Significance	105
Electrochemical Character and Electronic Configuration	105
Particle Size	107
Solubility of Metal Compounds and Hydration of Metal Ions	107
Hydrolysis and Olation	108
Colloidal and Radiocolloidal Behavior	110
Hard and Soft Acids and Bases Theory	111
Coordination and Chelation	115

CONTENTS

Protein–Metal Interactions 120
Nucleic Acid–Metal Interactions 123
Biologic Membrane–Metal Interactions 126

Chapter 5
CARCINOGENICITY AND TERATOGENICITY 129

Carcinogenicity 129
    Carcinogenic Metals 133
        Group I 133
        Group II 135
        Group III 137
        Group IV 137
        Group V 139
        Group VI 141
        Group VII 143
        Group VIII 143
    Mechanisms of Metal Carcinogenesis 145
    Surface Oncogenesis 148
Teratogenicity 153

Chapter 6
SUMMARY AND OVERVIEW OF METAL TOXICITY 161

Summary of Metal Toxicity by Group 164
    Group I Metals 164
        Subgroup IA Metals 165
        Subgroup IB Metals 166
    Group II Metals 167
        Subgroup IIA Metals 168
        Subgroup IIB Metals 169
    Group III Metals 171
        Subgroup IIIA Metals 172
        Subgroup IIIB Metals 173
    Group IV Metals 176
        Subgroup IVA Metals 176
        Subgroup IVB Metals 178
    Group V Metals 178
        Subgroup VA Metals 179
        Subgroup VB Metals 180

Group VI Metals 181
    Subgroup VIA Metals 181
    Subgroup VIB Metals 183
Group VII Metals 184
Group VIII Metals 185
Periodicity and Toxicity 188
    Chemical Toxicity on the Basis of Vertical Groups 188
    Chemical Toxicity on the Basis of Horizontal Periods 189
    Integration 191

APPENDIX A 193

APPENDIX B 197

GLOSSARY 201

REFERENCES 215

INDEX 233

1

INTRODUCTION TO HEAVY METAL TOXICITY IN MAMMALS

Throughout biologic evolution, living forms have incorporated into biochemical constituents those available metals that had physical and chemical properties suitable for carrying out the functions of life. The progress of evolution is a response to the ever-changing environment. Animal requirements for certain metals are a direct outgrowth of this association. Our knowledge of trace-metal requirements for mammalian growth, development, general health, and reproduction is growing rapidly as intense research progresses (Hopps, 1972). Certain metals that previously were considered to be toxicants are now considered essential nutrients. At the same time, increased concern and effort are extended to problems of metal toxicity and safety in man and other mammals. The actual and potential problems of metal toxicity are often examined with too-limited knowledge by environmentalists and by those in the news media who suggest when and where technology must stop in order to preserve the immediate environment of endangered species. Experiments readily show that any compound is harmful if given in excess! Science and technology allow man considerable control of his immediate environment, but the earth–inhabitant continuum cannot be denied. The secret of survival, therefore, is not to absolutely prohibit the use of any chemical, but to utilize each chemical rationally. The environmental quantities of all minerals and their salts must be maintained at levels compatible with continued optimum health and existence. We accept the meditation of Marcus Aurelius that where a man can live, there he can also live well.

The variation in concentrations of a given metal in different geologic settings is tremendous. Many minerals were not generally available to the biosphere until the action of water and other geologic forces distributed them from isolated caches to floodplains of rivers, alluvial deposits of glaciers, and selected waterways. The availability of the many metals in minerals depends on the same properties, such as solubility and reactivity, that make them useful in living systems. Certain environments develop unusually high concentrations of specific elements. Alkaline salts make some streams undrinkable. The oceans are so hypertonic that they are harmful to nonadapted mammals. The prairies of South Dakota have excess selenium which is incorporated into indigenous plants; the grazing of animals is considerably restricted by this excess. Indians and white settlers learned to keep their herds away from the luxuriant grasses in the low areas of the Dakotas to prevent alkali disease. Farmers must still exercise care to prevent excess selenium being fed to domestic animals.

Most environmental changes of the earth are slow enough to permit living forms to adapt to the changes. During the twentieth century, human activities have accelerated the rate of change in worldwide environments enough to tax the adaptability of living systems. With veins of coal for an umbilicus and with a gourmand appetite demanding ever more metal ores, civilization has developed gourmet needs for a buffet of rare metals in order to introduce the atomic age. In the past two decades, science, industry, medicine, and agriculture have exposed man and his world to ever-increasing numbers of exotic chemicals, represented by metal dusts and new organic chemicals. This activity exposed mammals to a wide array of metals that would otherwise have stayed buried or localized. Megalopolophilic humans accentuate potential and real problems of metal toxicity by their concentration of activity to restricted geographic areas via urbanization, agribusiness, and recreation.

For millenia, mankind ate what his forebears had eaten without difficulty. The dietary patterns of mammals are, to a large extent, learned habits reinforced by the occasional sickness or death of animals that ate harmful food. Knowledge of the relationship between food and harmful materials has developed scientifically in separate compartments as: (1) metals essential for mammals, (2) toxic metals, (3) radioactive metals, (4) stimulatory compounds, and (5) nonreactive or nontoxic metals. All of these now merge into one complex interaction between mammals and metals. Metals that were routinely discussed as toxic (selenium, nickel, vanadium, arsenic, and tin) are now considered possible essential nutritional elements, and the toxic limits of essential nutrients (cobalt, copper, chromium, iron, and zinc) are being seriously examined. Concepts of the interplay among metals as essential nutrients, stimulants, and toxicants provide greater truth and strength than do concepts derived from any part of these information bases.

Full information on the effects of different concentrations of any metal and its compounds and on both the synergistic and the antagonistic interactions of each metal with other compounds must be understood before a civilization can hope to survive our technologic age. These concepts, accepted by knowledgeable scientists, must be promulgated to the media, environmentalists, economists, forensic practitioners, and practical politicians for rational utilization of the chemicals in our lives.

Recognition that man or any given species is different from all other animals does not negate the general similarities present in the class Mammalia. Individual variation is anticipated in all biologic studies; the diversities found among species do not dispute the basic biochemical, morphologic, nutritional, and toxicologic unity of all mammalian species. Man is different from, but not separate from, other mammals. Data from all mammals were therefore utilized to develop general concepts of metal toxicity. This general understanding of toxicity will provide enlightened consideration of the probable activity of any metal in any given species under a given set of conditions, including an understanding of the potential for stimulation and nutritional requirements for metals and their inorganic salts and complexes. Generalizations from laboratory animal data are important for other mammals on which there is insufficient information. The basis for phylogenetic diversity developed from evolution as a two-compartment system: the developing genetic base and the ever-changing environment (Luckey, 1975). The role of intestinal microbes in animal evolution and development has been explored (Luckey, 1972), and the phylogenetic basis for differences in comparative animal toxicity has been summarized (Luckey, 1976a,b).

SCOPE AND LIMITATIONS

This treatise provides the basis for an understanding of the chemical toxicity of nonradioactive metals, metalloids, and their inorganic compounds. Literature on the toxicity of metals for mammals has been thoroughly reviewed, and pertinent information is summarized and interpreted. New concepts have been introduced. Classic compartments of knowledge have not been respected; for example, nutritional, stimulatory, therapeutic, and toxic effects of metals are presented as components of a single continuum of a complete dose–response concept. We emphasize that a knowledge of the complete biological activity spectrum of inorganic chemicals is useful and necessary in a technologic society. An ultraconservative approach to the use of chemicals in our environment is encouraged by a limited knowledge of toxicity and by the dramatic increase in sensitivity of analytical methods.

Understanding of the physiologic and chemical basis of metal toxicity in mammals will be useful as a guide to present practice and to future research, legislation, and regulation. With this objective, we have suggested some generalizations from available toxicity data. In the absence of specific toxicity data about some less-common metals, these generalizations should provide: (1) the base for predictive toxicity, (2) guidelines for safety, and (3) the motivation to obtain more complete toxicity data. In these generalizations, each species of animal is considered as distinct within, but not separate from, the class Mammalia. Man is no exception. We have tried to draw all information possible from the limited data available, especially for some of the less-common metals. Early experimental and clinical data on the toxicity of metals may be unreliable due to poor analytic methods and/or the presence in trace amounts of contaminating metals in the materials tested. Some experiments had either inadequate or no control groups. These generalizations should therefore be modified as more data are obtained.

The state of the subjects has been given little consideration; most of the animal work assumes an ideal, caged, well-fed subject. This ignores the multiple stresses of everyday life and the extent to which environment, diet, stresses, disease and repair processes, and therapy all influence toxicity. Diurnal variations are significant (Halberg, 1969).

Since most of the infectious diseases have been effectively controlled, degenerative diseases have become of greater concern. Environmental pollution is implicated in these degenerative diseases. Since metals and their compounds are major components of pollution, their role in carcinogenicity and teratogenicity, including the phenomenon of surface carcinogenesis, is reviewed in this volume.

The geologic occurrence, biologic action, industrial uses, and the physical and chemical properties of each metal are summarized. What are the basic reactions of the metal salts in physiologic milieu? How much of a given material might be absorbed by different routes? How much of what is absorbed might be retained in tissues? How does the body excrete or detoxify metals? Is there an effective physiologic homeostasis? To the extent of available information, these questions are answered for each metal and each group of metals. A glossary is provided in the appendix.

In order to understand the basic reactions possible between metals and other biochemical constituents, the hard and soft acid and base theory of Pearson (1963) is reviewed. This leads to a better understanding of coordination, chelation, ligand formation, and metal interaction with proteins, nucleic acids, carbohydrates, and lipoprotein of cell membranes and organelles. A comparison of the physicochemical and metabolic characteristics of a metal with those of neighbor toxic metals enables some prediction of the toxicity of the related metal. Such extrapolation holds true in some cases. However, the unexpected property of thallium, a Group III metal, to form stable Tl^+ salts, is more important in its toxicologic reactions than the usual characteristics of Group III metals in the periodic table.

This treatise is a modest contribution to the biological sciences. Major limitations are obvious; it is not meant exclusively for toxicologists, pharmacologists, nutritionists, biochemists, ecologists, physicians, or veterinarians. It is meant for all. Detailed medical viewpoints, diagnoses, pathologies, and treatments of metal toxicants are generally not given. We have found no systematic presentations for discussion of medical or metabolic reactions. Nonmetallic elements are not considered. Radiation toxicity is excluded, and the chemical toxicity of highly radioactive metals is briefly reviewed as needed. Although simple organic metal salts have been included, the vast array of organometallic compounds has been ignored; the study of these should comprise a third volume. Generally, the simple organic metals are the most toxic (as exemplified by metal carbonyls), while many complex organic metals are tightly bound into un-ionized and unavailable forms that are relatively harmless. Our reviews of the toxicology of lead, mercury, and cadmium give relatively much less information than do recent books written about each of these metals.

This volume provides introductory and background information about metals and their salts in mammalian nutrition, physiology, and toxicology. New concepts to evaluate safety and to express toxicity are developed. The volume ends with a summary of metal toxicity in mammals, using data from Volume 2. The second volume presents a thorough review of the literature of the toxicity of all nonradioactive metals and metalloids. The extensive references of Volume 2 include general reviews for each metal wherever possible. The references to and information about little-discussed but highly toxic metals such as indium and thallium are an alert to these dangerous components in our environment. The complete work should allow the reader to evaluate the toxicity, safety, and utility of metals for their judicious use in a technologic society.

ESSENTIAL METALS

Philosophically, organisms develop through nurture of their nature. *Nurture* refers to those environmental factors that contribute to development; *nature* is defined as the genetic potential of an individual through time. Nutrition is the action of the chemical component of the total environment on living organisms; it includes harmful as well as essential chemicals eaten, drunk, injected, aspirated, or absorbed through the skin (Fig. 1-1). The complete dose–response curve for essential metals shows a general complex activity spectrum comparable to that of Fig. 1-2. The generalized concept of nutrition (Fig. 1-1) suggests that nutrients, stimulants, and toxicants are all parts of a single component–the chemical environment. Arsenic, a classic poison and a general stimulant, is now being investigated as an essential element. Classically, cadmium, selenium, silicon, and tin have long histories as toxic elements; in the past decade, each has been proposed as an essential nutrient for animals. Alternately, many essential metals are being investigated for their toxicity in excessive amounts.

The proposal for any new dietary essential metal prompts reexamination of the criteria for an essential nutrient (Luckey, 1976^a). What is the critical distinction among essential nutrient, stimulant and therapeutic agent? The criteria and the evolutionary base for essential nutrients have been discussed. An essential nutrient is usually designated by the name of the most active of a family of active compounds; however, for minerals, each element represents a single entity. Salts and/or anionic forms of certain amphoteric metals should show high specificity to prevent or alleviate specific deficiency symptoms. Taking these symptoms as criteria, the credibility of a metal as an essential nutrient increases with alleviation of each of the following criteria, particularly if it is unaffected by closely

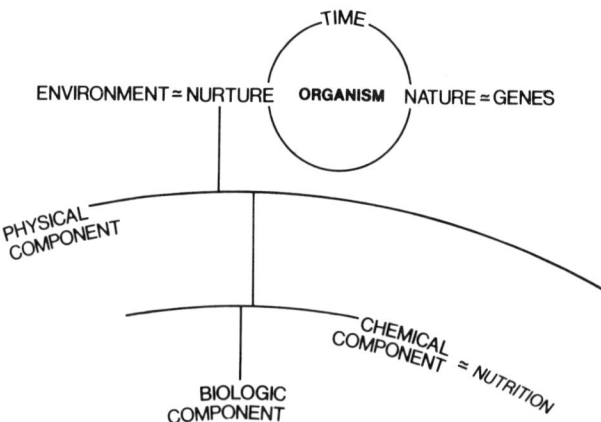

FIG. 1-1. Philosophic definition of nutrition as the chemical component of nurture. Note that this definition includes essential nutrients, nonessential nutrients, nonnutrients, hormetins, and toxic materials.

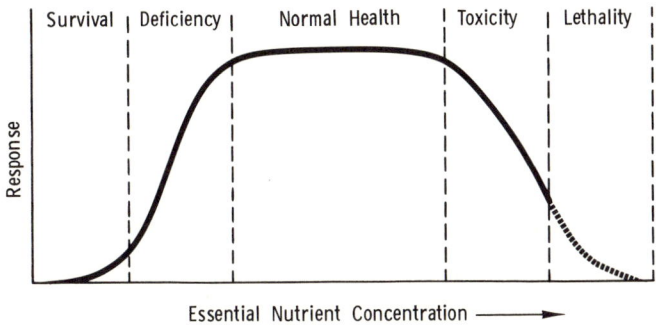

FIG. 1-2. Activity spectrum of an essential mineral.

related elements: (1) decreased appetite, (2) increased requirement for another nutrient, (3) biochemically significant change in tissue composition, (4) decreased growth or increased morbidity and/or mortality under special conditions, (5) disability under special conditions, (6) subclinical disability, (7) generally increased morbidity and/or clinical disability, (8) retarded growth, (9) inability to function normally (e.g., reproduction), and (10) death. Since the criteria are arranged in order of increasing importance, a rough estimate of credibility as an essential nutrient may be obtained by adding the numbers of those criteria for which any nutrient has shown specificity. The values obtained would be high for classic essential nutrients such as iron; they would be minimal for lithium, more for strontium, and higher for nickel or tin. Appropriate credibility values are obtained for all established essential minerals for vertebrates (Fig. 1-3). This subject has been well reviewed (Underwood, 1971; Krook, 1976; Williams, 1976; Schwarz, 1972).

All the minerals noted as "established" in Fig. 1-3 are accepted as being required by all vertebrates. Mammals have no specific requirements for minerals that are not required generally by other animals. Some elements, formerly known to be functional requirements for lower vertebrates (i.e., vanadium) or intervertebrates (i.e., silicon), are being suggested as essential nutrients for mammals. Research with invertebrates and/or *in vitro* studies with vertebrate tissues provide evidence that certain other elements may become accepted as essential nutrients. These and elements required only under special conditions (e.g., ultraclean environments) are indicated by incomplete boxes in Fig. 1-3. All the elements marked by partial boxes and certain other metals are essential to plants and/or microbes. The number of metals considered as essential for mammals is increasing and may continue to increase due to the intensive research

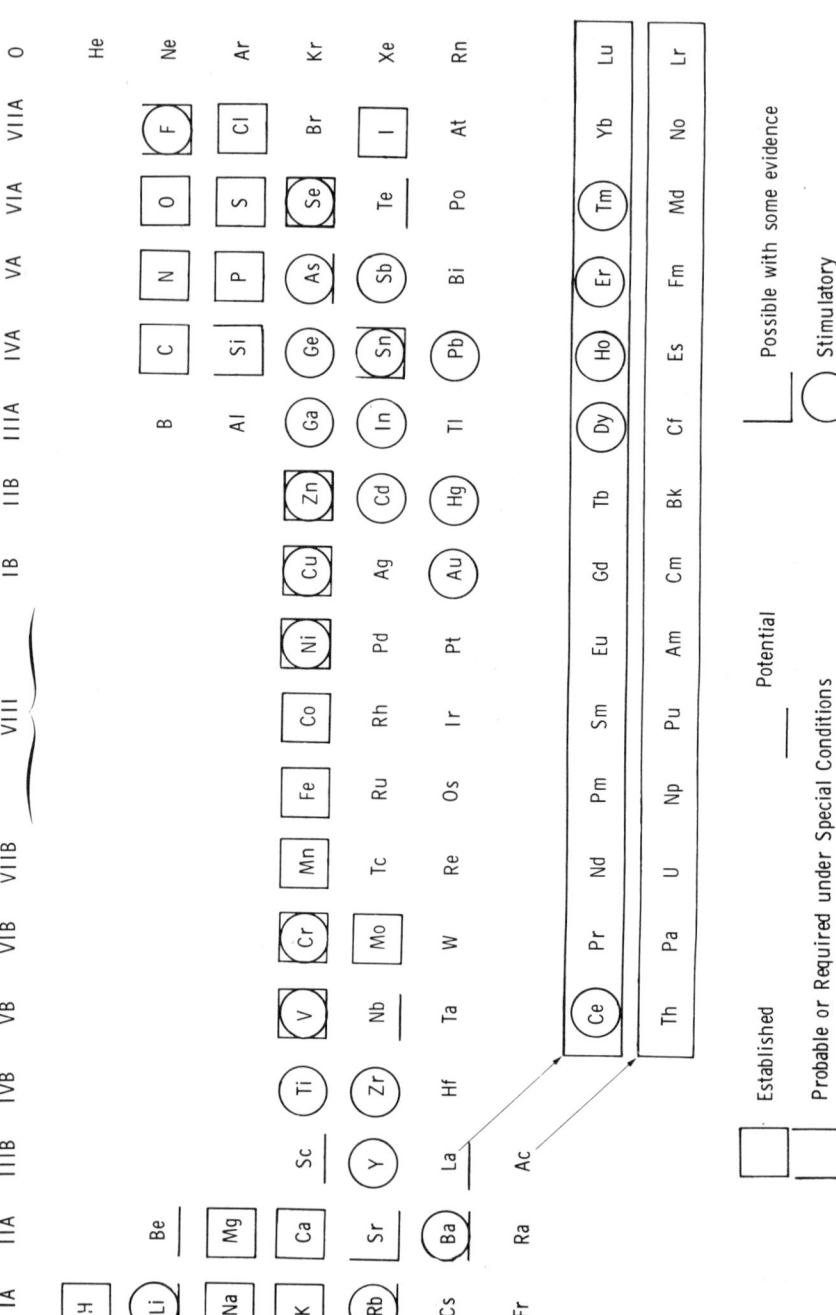

FIG. 1-3. Categorization of elements of the atomic table from the viewpoint of nutrition and hormology.

during the present decade. Few would rate high by the criteria above, and there is a real need to distinguish new essential metals, such as selenium, from stimulatory metals, such as germanium.

STIMULATORY METALS

The confidence with which an element is considered to be essential may be shaken when one studies growth promotants and dietary therapeutics. Dose–response curves comparable to that of Fig. 1-2 are frequently noted when nonessential metals are administered to animals (Luckey, 1975). Elements in every group of the periodic table have been found to be stimulatory to animals (Fig. 1-3). It is estimated that 50% of all potentially harmful agents will be stimulatory in minute doses. Such compounds are known as *hormetins*. Appreciation of such a complex dose–response curve is vital to a full evaluation of the range and ultimate usefulness of hormology, the study of excitation. The biologic activity of low doses of metal compounds must be known in order to understand the effect of metals in living systems. A general view of the different types of hormetic action (Fig. 1-4) shows that the response discussed above is the β type. The α curve shows no stimulation; appropriately small doses of α type hormetics may never have been examined, or some compounds may exert no hormesis. The γ and δ curves have not been studied extensively. During a thorough study of the oligodynamic action of heavy metals in bacteria, Richet (1906) found each type of stimulation (Fig. 1-5). That none of his compounds showed simple toxicity, the α dose–response curve, suggests that the predominance of this curve noted in mammalian toxicology reflects an incomplete examination, with minute doses, of the "toxic material." Well-known examples of the complex curve are the use of high levels of copper in pigs and chickens; this stimulatory action of copper is achieved only with dosages about 10 times greater than its requirement (Barber *et al.*, 1956; Braude, 1965). As illustrated in Fig. 1-6, a hormetic may also be an essential nutrient. Growth stimulation by a nutrient at a level below the nutrient requirement was noted for sodium in crickets (Luckey and Stone, 1960), but has not been seen in mammals. The hormetic action of a compound is generally observed at 10–1000 times less than the toxic dose. The stimulatory dose may be difficult to find if it is a sharp peak with a narrow concentration range. Few researchers examine the complete dose–response spectra.

The general usefulness of these concepts of hormology is in replacing the simplistic labeling of a compound as good or bad with more accurate

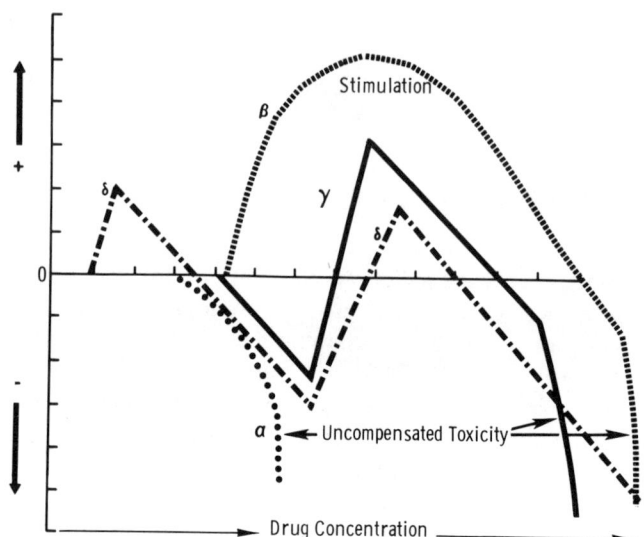

FIG. 1-4. Three types of reactions observed in hormology. The ordinate indicates the response as change from control. the β curve is the only one that has been examined experimentally.

descriptions of the activity of any chemical depending on the dosage and conditions. A specific use for this part of the dose–response curve is to indicate a concentration that is biologically equivalent to zero for the parameter(s) measured. Since most toxicants are stimulatory when appropriate low doses are administered, they provide an objective zero equivalent point (ZEP), which should be used to replace zero tolerance for harmful compounds by legislative and regulatory organizations. The ZEP is the highest quantity of the agent that gives results not detectably different from controls given placebos.

PHARMACOLOGIC AND THERAPEUTIC USE OF METALS

Before the advent of organic antibiotic and bactericidal agents, inorganic metal salts were used extensively as therapeutic agents. Salts such as potassium permanganate are presently used in dilute aqueous solutions as antiseptics. Some metal salts combine with proteins and organic acids to act as local astringents, irritants, and antiseptics. Continuous exposure to small quantities of metals produces cumulative effects that may result in

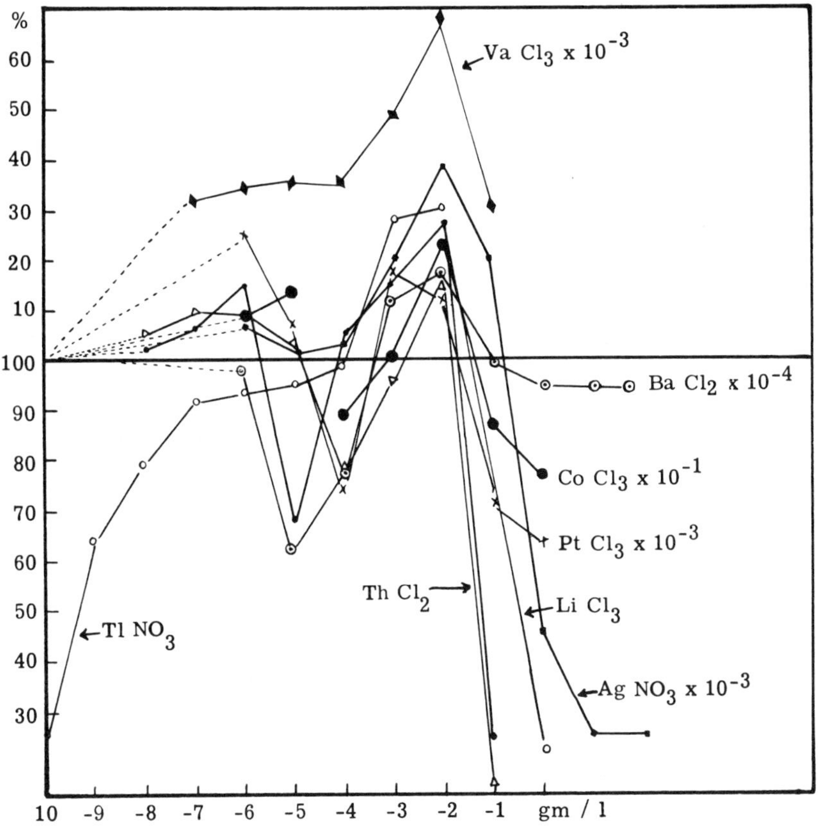

INTRODUCTION TO HEAVY METAL TOXICITY IN MAMMALS

FIG. 1-5. Fermentation stimulation by metals (Richet, 1906).

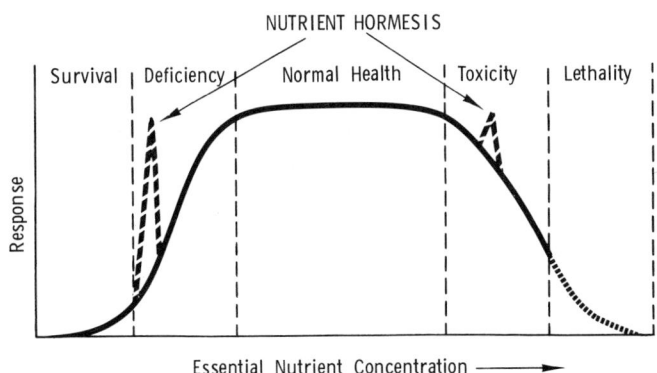

FIG. 1-6. The hormetic action of nutrients.

chronic poisoning with metabolic, nutritional, and neurologic symptoms. In larger doses, these salts become acutely toxic, the toxicity depending on the physiochemical properties of the salt. Many quantitative data on metal toxicity were derived from therapeutic uses of metal compounds. For example, a variety of organic arsenic compounds and mercury compounds were used as chemotherapeutic agents against parasitic infection. When the active salts showed no, or relatively low, toxic effects in host organisms, these compounds were said to have a favorable chemotherapeutic index; i.e., a low ratio of minimal effective dose/minimal lethal dose. A summary of therapeutic uses of metals follows (Merck and Co., 1968).

Among Group IV metals, germanium, tin, lead, titanium, and zircomanic psychoses. Apart from their essential nutritional value, copper salts were used therapeutically as stimulants and as astringents in conjunctivitis, urethritis, and vaginitis; their high toxicity for fungi, protozoa, and algae has been utilized for water purification. Colloidal silver salts, especially the argyrol type (mild protein silver), act as antiseptics on mucous membranes. In traces, ionic silver is antiseptic; it acts as a mucilaginous demulcent and detergent to dislodge pus. Gold compounds such as gold thiomalate given intramuscularly are effective in treatment of rheumatoid arthritis, especially in early affliction; they apparently arrest the inflammatory process. The side effects of this therapy are unpredictable and may be undesirable.

Among Group II metals, magnesium, calcium, barium, zinc, and mercury are used therapeutically. Magnesium sulfate is an efficient osmotic cathartic, magnesium oxide and magnesium carbonate are efficient antacids, and their antacid action extends into the intestines with little interference with stomach physiology. The insoluble barium sulfate is used in X-ray diagnosis, since it is opaque and interferes little with stomach movements. Large quantities may act as a mild cathartic. The therapeutic uses of calcium salts are numerous.

Zinc oxide and salts such as acetate, subcarbonate, nitrate, and sulfate are used as mild astringents and antiseptics; zinc sulfate is a mild emetic. Zinc undecolinate is commonly used to treat "athlete's foot."

Toxic mercuric salts, such as chloride, iodide, oxide, and salicylate, are used as antiseptics, disinfectants, and fumigants; mercury oleate is a topical parasiticide. Mercurous chloride was used formerly as a cathartic, diuretic, antiseptic, and antisyphilitic.

Among Group III metals, salts of aluminum, indium, and lanthanides were used therapeutically. Some of the soluble aluminum salts were used as antiseptics, astringents, and styptics; aluminum hydroxide is a good gastric antacid. Radioactive indium complexes (organic) are utilized to locate tumors. Lanthanide salts have been used as anticoagulants.

Among Group IV metals, germanium, tin, lead, titanium, and zirconium salts were used therapeutically; germanium oxide was formerly used as a bone marrow stimulant in anemia. Lead acetate was used as a topical astringent. Stannous fluoride is used topically for prevention of dental caries. Titanium sulfate and dioxide were used topically for skin diseases, rhinitis, and conjunctivitis. Zirconium oxide ointments are used for poison ivy dermatitis.

In Group V, inorganic arsenic compounds were formerly used extensively and somewhat indiscriminately in nutritional diseases (including anemia and pellagra), asthma, rheumatism, neuralgia, cholera, malaria, skin diseases, and syphilis. Inorganic arsenic is restricted in the United States to treatment of certain skin diseases and leukemia (Merck and Co., 1968). Intravenous injections of potassium antimony tartrate (tartar emetic) are highly effective against a number of tropical diseases due to animal parasites: kala-azar, filariasis, leishmaniasis. It is now being replaced by organic compounds. Organic metallic compounds are used more and more. Bismuth subcarbonate is a slow-acting antacid and a most effective nonirritant antiseptic; it allays diarrhea. Bismuth salts are effective against syphilis and trypanosomes, but have toxic side effects on the host organisms. Vanadium has spirocheticidal action and was formerly used therapeutically.

Among Group VI metals, chromium salts were used therapeutically; chromium oxide is a topical astringent, and labeled sodium chromate is an erythrocyte-tracing agent. Sodium selenite is used in veterinary medicine.

Among Group VII metals, $KMnO_4$ is used as an antiseptic, and manganese carbonate and manganese hypophosphite were formerly used as hematinics; labeled technetium salts are used in organ-scanning, especially brain and thyroid.

In group VIII salts, such as ferric ammonium sulfate, pyrophosphate, ferrous citrate, and cobalt chloride, are used as hematinics. Ferrous iodide was used in the treatment of chronic tuberculosis, and ferric chloride is used as a local astringent and styptic.

The inorganic salts described above are used therapeutically for humans and other mammals in some areas of the world despite the risks involved.

TOXICOLOGY

Toxicology is the study of the adverse and harmful effects of chemical agents on any biologic system; the word *toxic* is derived from the Greek words *toxon* ("bow") and *toxicon* or *pharmikon* ("arrow poison"). Toxicology embraces other scientific disciplines such as chemistry, biology, physiology, nutrition, pharmacology, pathology, immunology, and medicine. Toxicology comprises many areas of service and research. Environmental toxicology is that branch of toxicology that deals with the incidental exposure of man and other animals to harmful contaminants of the environment. Forensic toxicology deals with the medical and legal aspects of the adverse effects of chemicals on humans. Behavioral toxicology is a new concept (Mello, 1975; Spyker, 1975). Clinical toxicology deals with the study of the diagnosis and treatment of diseases resulting from adverse effects of chemicals. Experimental toxicology studies the effects of toxic levels of chemicals and drugs. Industrial toxicology deals with material involved in occupational health hazards and industrial hygiene. Economic toxicology is the study of such agents as insecticides, herbicides, and defoliants and their effect on pests, domestic animals, and humans.

Modern industrialization has introduced harmful metals into the environment by redistributing them from immobilized ores and minerals. Metal salts have been used therapeutically in human and veterinary medicine for centuries, but man and animals are now exposed to more metal salts than they were previously. Metal compounds used in pesticides, raw materials, catalysts, or energy end up as residues in food, water, and air. Metals for which there are no nutritional requirements may react with biological systems to cause adverse effects. Excessive doses of nutritionally essential metals can also cause adverse effects. Excessive absorption and/or renal insufficiency or biliary obstruction lead to the breakdown of homeostatic mechanisms and to the accumulation of metals in tissue levels high enough to cause toxic effects.

The toxicity of a metal depends on the inherent capacity of a material to affect adversely any biologic activity. *Toxicity* is a relative term used to compare one chemical or metallic compound with another. Essential metals maintain dynamic equilibria in life processes. Toxic metals (including excessive levels of essential metals) tend to change biologic structures and systems into irreversible and inflexible conformations leading to deformity or death. The generalized response of an organism to a toxicant is biphasic, a phase of biologic effects being followed by a phase of pharmacotoxic actions.

Toxicity is only a single component in the complete activity spectrum or dose–response curve of a chemical in a living biologic system (see Fig. 1-2). If the metal is not essential, the two categories on the left (survival and deficiency) are not germane. Depending on its characteristics and dose, a metal may be innocuous, essential as a nutrient, stimulatory, therapeutic, harmful, or lethal. The response varies according to the environment, diet, and condition of the animal, as well as the form, dosage, and mode of

INTRODUCTION TO HEAVY METAL TOXICITY IN MAMMALS

administration of the metal or its salt. Synergisms and antagonisms among metals are included in these factors. Only a broad overview will allow a thorough understanding of the complex activity spectrum of metals and their compounds. This understanding is needed to help counteract the simplistic labeling of a given metal salt or a chemical as "good" or "bad," and to enable society to make enlightened decisions on its biologic functions and limitations.

Criteria for metal toxicity in mammals are early mortality, growth retardation, impaired reproduction with mortality of offspring, depression of physiologic parameters, neoplasms, and chronic disease symptoms. At the cellular level, derangement of cell-membrane permeability and antimetabolite activity are the effects of metal toxicity. Metals can interact with a protein, leading to an allosteric effect, or with DNA or RNA to stop normal metabolism, or with unknown compounds, leading to a change in physiologic processes, to a change in behavior, or even to a change in an ecological system. More than one system is affected by each toxicant. Changes in rates of the catalytic decomposition of essential metabolites, enzyme inhibition, and irreversible conformational changes in macromolecular structure are some of the effects of metal toxicity at the molecular level. Biological and environmental variations influence the toxicity of metals. The inherent toxicity of a metal and its compounds in biologic systems depends on the following: its electrochemical character and oxidation state, its absorption and transport in the body tissues, the stability and solubility of its compounds in body fluids, its ease of excretion, and its reaction with functioning tissues and organelles and with essential metabolites and other metals.

Toxic effects are dose-dependent; the effects of a metal can be categorized as doses that cause: (1) no symptoms or detectable effects, (2) stimulatory effects, (3) therapeutic effects, (4) toxic or harmful effects, and (5) death.

Acute or proximal toxicosis (Pham-Huu-Chanh, 1965) is caused by a relatively large dose of a metallic toxicant; the onset of symptoms is sudden, and the intensity of effects rises rapidly and may result in death. If adequate procedures are not performed to neutralize or remove the toxicant, irreversible damage to tissues and systems may cause death. Chronic poisoning, also called cumulative poisoning or distal toxicity, develops gradually following long and continued exposure to relatively small doses. There is a period with no symptoms, followed by a gradual onset of symptoms. There may be frequent remissions and recurrences of the symptoms. Many metals act as short-term poisons or toxicants in high doses and as long-term systemic poisons in low doses. Chronic poisoning also represents cumulative effects. A metallic toxicant such as mercury can develop two different sets of symptoms, one for chronic and one for acute toxicosis. Acute mercury toxicosis from oral ingestion of mercuric chloride causes a severe course of nausea, vomiting, and diarrhea culminating in death due to circulatory failure. Chronic mercury toxicosis causes progressive renal failure with circulatory and neuromuscular abnormalities. Chronic toxicosis can be reversed by removing the toxicants, provided no irreversible damage has been done to vital systems. Removal from further exposure and the restorative potential of the body both hasten recovery from the chronic toxicosis. Schroeder (1973) explored "recondite toxicity"; rodents developed symptoms after only one to three generations of drinking water with small quantities of metal salts. Groth (1972) found that metal interactions were effective in chronic experiments, with mercury toxicity being alleviated by selenium. Delayed, or latent, toxicosis is the condition in which clinical effects are observable only months following exposure to the toxicant. Latent toxicosis is exemplified by beryllium and chromium. Cumulative effects may or may not be associated with a buildup of the toxicant; there is no cumulative concentration of beryllium in pulmonary tissues in the pulmonary granuloma caused by chronic beryllium toxicity.

Tolerance is the ability to endure the continued and/or increasing administration of a toxicant. Tolerance is also the capacity of an organism to exhibit less response to a test dose of a chemical than it did previously to the same dose. Rodents exhibit this tendency to a dose of chloroplatinates. The same phenomenon is observed when a large single dose is administered a few days after the administration of a number of small doses of the same compound. The organism either had become refractory or had developed resistance to the toxicant. Tolerance should not be confused with one of its components, resistance, the capacity that allows the animal to inactivate or metabolize the toxicant.

Quantification of the dose–response relationship leads to standard values of assessing the comparative toxicity of toxicants. Trevan (1927) introduced the term LD_{50}, which is defined as the minimum dose of a toxicant at which 50% of a population of test animals will die. The LD_{50} value is a statistically obtained value; it is interpolated from the sigmoidal dose–response curve, which was drawn by plotting percentage mortality at each dose against the dose. This graphic method has been modified to suit both small and large populations of test animals. The method of Litchfield and Wilcoxon (1949) is the most versatile and is commonly used for a large number of test animals in determination of LD_{50} values with confidence limits. The method of Behrens and Kärber (1935) is used for smaller numbers of test animals. The logarithmic probit graph paper method of Miller and Tainter (1944) and the range-finding procedure of Weil (1952) can also be used to interpolate the LD_{50} value. The LD_{100} of a toxicant is the dose at which all the test animals die, while the minimum lethal dose (MLD) represents the dose that will kill at least 1 animal in the test population. This dose is interpreted here to be LD_1 for purposes of our discussion. The doses are usually expressed as milligrams of the metal or salt per kilogram body weight of the animal. Expressing these values on a molal basis (mol/kg) provides a better index for comparing the innate toxicities of similar metals, because metals that differ little in their atomic weights may have great differences in their specific gravity. These values should be qualified, however, by specifying chemical form of the metal, mode of intake, condition and developmental stage of test animals, and the time between the administration of the toxicant to the animal and its death.

Available toxicity data on metals often do not provide the time vector in LD experiments. The interpretation of the available LD_{50} values is nebulous for evaluation of distal toxicity and for latent toxicity. In distal toxicity, the LD_{50} diminishes with the length of the observation period and becomes constant at the end of this period, called the *period of crisis*. If the LD_{50} value remains unchanged whatever the duration of the observation period, it is called the *aggression index*. It is difficult to assess the inherent toxicity of a metal in distal toxicity on the basis of the available LD_{50} data without time being stated. To avoid the anomalies described above, new terms were introduced by Pham-Huu-Chanh (1965) to express lethal doses: (1) maximum dose never fatal in 24 hr (MDNF), and (2) minimum dose always fatal (MDAF) (smallest dose causing 100% mortality in 24 hr). So far, few data have been accumulated expressing toxicity values as MDNF and MDAF. Comparable data on MDAF values for sodium chromate, molybdate, and tungstate (metals of Group VIB) expressed on mass and molar bases (Table 1-1) justify and reiterate the necessity for expressing toxicity data on a molal basis with time as a vector. The specific and innate toxicity of certain metals is best expressed by the ratio of beneficial to toxic dose: The lower the ratio, the greater the toxicity. On this basis, hexavalent chromium with a ratio of 1:100 is a nontoxic metal ion relative to selenium or arsenic salts; these criteria, however, give no hint that chromium is carcinogenic.

Expression of data in inhalation toxicity as parts per million of the toxicant metal per cubic meter of air does not reflect the correct and absolute dose of the toxicant inhaled by the test animal. At best, these values depict only threshold values. Small aerosol particles with a diameter of less that 5 μm reach the alveoli of the lungs; the bigger particles are retained in the nasopharyngeal region and are expelled or deposited into the gastrointestinal tract. Rigid experimental conditions, including time and particle size and mass distribution by particle size, should be stated to

TABLE 1-1. Comparison of Toxicity Values of Chromate, Molybdate, and Tungstate on Mass and Molar Bases

Sodium salt	Specific gravity		Molecular weight of compound	Atomic weight of metal	LD_{100} or minimum dose always fatal (MDAF)[a,b]		
	Metal	Compound			mg Compound[c]	mg Metal[c]	mmol Compound[c]
Chromate	7.2	2.71	161.97	51.99	121.5	39	0.75
Molybdate	10.2	3.28	205.92	95.95	721	336	3.5
Tungstate	19.35	4.18	293.83	183.85	293	184	1.0

[a]MDAF values in mmol compound/kg body weight (from Pham-Huu-Chanh, 1965).
[b]By intraperitoneal injection in rats.
[c]Values expressed per kilogram body weight.

TABLE 1-2. Classification of Metal Toxicity[a]

	Oral[b]	Intravenous[b]	Inhalation[c]
Relatively harmless	15,000	1,500	100,000
Practically nontoxic	5,000–15,000	500–1,500	10,000–100,000
Slightly toxic	500–5,000	50–500	1,000–10,000
Moderately toxic	50–500	5–50	100–1,000
Highly toxic	5–50	0.5–5	10–100
Extremely toxic	1–5	0.1–0.5	5–10
Super toxic	<1	<0.1	<5

[a]Adapted from Hodge and Sterner (1943).
[b]Single dose expressed as mg metal/kg body weight.
[c]Exposure of 4 hr gave 33–66% mortality; number of particles/m^3.

express data in inhalation toxicity. Correlation of aerosol concentration to mg inhaled is needed.

Despite these limitations on the usefulness of the available LD_{50} data, they were used to classify the metallic toxicants into seven groups, adopting the classification of Hodge and Sterner (1943). The classification is shown in Table 1-2.

A new view of comparative metal toxicity is provided by emulating the pH concept: $pT = -\log [T]$, where pT is the potential toxicity, $[T]$ is the molal concentration of the compound or the metal, or mole/kg body weight, and the logarithm is to the base 10 (Luckey and Venugopal, 1977). A material that was found to give an LD_{50} at a concentration of 0.0001 mol or equivalent weight would be expressed as $[T]_{50} = 10^{-4}$ mol and $pT = 4$. Sample calculations for pT from concentration data and concentration from pT values are given in Appendix B. The relationship between pT and the molal concentration of a toxicant can be readily envisioned from examples based on a single LD_{50} dose (Table 1-3). The pT class of any material is defined to be the whole number that is used to define the specific pT value expressed. Thus, a compound with a pT_{50} value of 1.96 would be placed in class 1; a compound with a pT value of 2.02 would be placed in class 2. The species, mode of administration, and time of observation should be designated. Although *molal* is technically correct, *molar* is used routinely; they are equivalent terms for most animal toxicity work.

The disadvantages of introducing a new concept are overcome by the simplicity of the final expression and the fact that biologic toxicity is related to the dose on a logarithmic basis. Since pT expresses the power of the toxicant, it corresponds directly to the effect of the material on a given population. A small pT indicates a relatively harmless material, while a large pT indicates a highly toxic material. Materials or metals that were

TABLE 1-3. **Comparison of Expressions for Toxicants**

Description[a]	Oral toxicity (LD_{50})		
	mg/kg	mol/kg[b]	pT_{50}
Supertoxic	0.1	0.000001	6
Extremely toxic	1	0.00001	5
Highly toxic	10	0.0001	4
Moderately toxic	100	0.001	3
Slightly toxic	1,000	0.01	2
Practically nontoxic	10,000	0.1	1
Relatively harmless	100,000	1	0
Harmless	1,000,000	10	−1

[a] These descriptions are presented from past experience; they are not adequate for describing toxins.
[b] Assume a molecular weight of 100 amu for comparative purposes.

harmful in greater than molal quantities would be indicated by a negative pT value. Comparisons of the different categories are provided in Table 1-3 for oral LD_{50}. Similar comparisons could be made for MLD, LD_{75}, and LD_{100}, using pT_1, pT_{75}, and pT_{100}, respectively. We should accept molal concentrations as being most meaningful for comparative purposes; density, molecular weight, and atomic weight are all variables in the practical and useful "milligrams compound per kilogram body weight." The wide range of values obtained for the toxicity of any group of substances suggests that a simplification would be useful.

Comparative LD_{50} (ip) toxicity data on some metallic toxicants, plant and animal toxicants, and bacterial and fungal toxins are tabulated as mg/kg, mmol/kg, and pT (Table 1-4). All toxic compounds can be classified into 16 groups on the basis of pT_{50} values, with the ultratoxic botulinal toxin D having the highest value of 15.5. Most metallic toxicants by oral route would occupy pT classes 1–4. The data of literature are expressed by both standard and pT classifications throughout Volume 2, and a summary of all values is provided in Chapter 6 of this volume.

SAFETY

Safety and *nontoxic* are elusive terms in toxicology. Every chemical agent is inherently harmful, depending on form, dose administered, mode of entry, and character of the organism. "Nontoxic" generally means a low toxicity hazard by the oral route. Minute doses of every nonradioactive metal or metal salt can be taken safely by a mammal. Excepting allergic reactions, the body can tolerate a normal exposure via ingestion, inhalation, or dermal contact to low doses without developing any observable undesired effect. The safety limits of small doses of metal salts are defined by the inherent characteristics of the compound, homeostasis, interaction with other metals, simple dilution and distribution in the body mass, and inability to reach susceptible tissues at sufficient levels to cause damage. Thus, any dose that produces no effect can be considered to be a safe dose

TABLE 1-4. Mouse $LD_{50/ip}$ Data Classified by pT

pT class	Toxin or toxicant	Molecular weight	Toxicity mg/kg	Toxicity mol/kg	pT_{50}
15	Botulinal D	1,000,000	3.2×10^{-7}	3.20×10^{-16}	15.49
14	Botulinal A	900,000	1.14×10^{-6}	1.27×10^{-15}	14.90
13	Tetanus	66,000	1.67×10^{-6}	2.53×10^{-14}	13.60
11	Botulinal E	350,000	2.50×10^{-3}	7.14×10^{-12}	11.15
10	Palytoxin	3,300	1.5×10^{-4}	4.55×10^{-11}	10.34
9	Perfringens CΘ	74,000	8.1×10^{-3}	4.39×10^{-10}	9.36
8	Saxitoxin	372	3.4×10^{-3}	9.14×10^{-9}	8.04
7	Tetrodotoxin	319.3	1×10^{-2}	3.13×10^{-8}	7.50
6	α-Amanitin	916	0.3	3.28×10^{-7}	6.48
	Actinomycin D	1,255	0.7	5.58×10^{-7}	6.25
5	Strychnine	334.4	0.98	2.93×10^{-6}	5.53
	Rotenone	394.5	2.8	7.10×10^{-6}	5.15
4	$HgCl_2$	271.5	5	1.84×10^{-5}	4.74
	Parathion	291.3	5.5	1.89×10^{-5}	4.72
	Epinephrine	183.3	4	2.18×10^{-5}	4.66
	NaH_2AsO_4	163.9	9	5.49×10^{-5}	4.26
	TlCl	239.8	24	1.00×10^{-4}	4.00
3	HCN	27.0	3	1.11×10^{-4}	3.95
	$BeCl_2$	79.9	12	1.50×10^{-4}	3.82
	Na fluoroacetate	100.1	18	1.80×10^{-4}	3.74
	Morphine	285.4	285	9.99×10^{-4}	3.00
2	Streptomycin	581.6	610	1.05×10^{-3}	2.98
	Caffeine	194.2	250	1.29×10^{-3}	2.89
	Tetracycline HCl	480.9	650	1.35×10^{-3}	2.87
	$BaCl_2$	208.3	500	2.40×10^{-3}	2.62
	Aspirin	180.2	495	2.75×10^{-3}	2.56
	NaF	42	125	2.98×10^{-3}	2.53
	Pu citrate	435.1	1750	4.02×10^{-3}	2.40
	$CaCl_2$	183.3	1350	7.36×10^{-3}	2.13
1	Nicotinic acid	123.1	1860	1.51×10^{-2}	1.82
	CCl_4	153.8	4620	3.00×10^{-2}	1.52
	NaCl	58.4	2600	4.45×10^{-2}	1.35

Data taken from Luckey and Venugopal (1977).

of that chemical for that specific organism or species under the conditions examined. This concept is the basis for zero tolerance introduced by the legislators for pesticide and insecticide residues in foodstuffs.

Toxicity studies that examine a wide range of doses of a toxic compound would allow extrapolation to a negligible-effect dose in experimental animals as compared with control animals given a placebo. This quantity delineates the zero equivalent point (ZEP) on the dose–response curve (Luckey, 1975). Inhalation studies recognize ZEP as the "threshold value." Many metal salts will be stimulatory at doses lower than the ZEP (Fig. 1-7). Thus, dose–response curves showing only toxicity should not be extrapolated directly to zero. A ZEP or safe dose can be determined experimentally for specific parameters. It should be obtainable from either hormetic or toxicologic data. The route of administration, diet, and other conditions are significant and should be specified.

Criteria of safety must be clearly set forth. These criteria may differ under pathologic conditions and under conditions of health, and these changes assume greater significance in humans. The safety limits of a chemical determined for two species of animals, rodents and nonrodents, may be applicable to other species and may form a basis for limited application to humans (Boyland, 1968).

Although an approximation to ZEP may be made from either toxicity

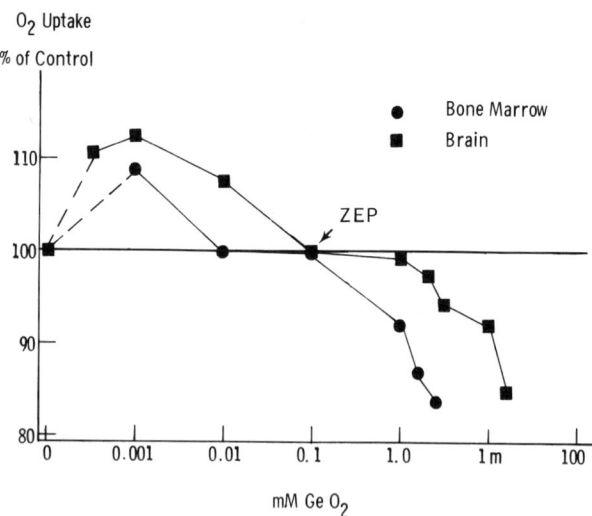

FIG. 1-7. Zero equivalent point (ZEP), defined experimentally as exemplified by germanium oxide. Data from Read (1949). Similar data have been obtained with chick growth (E. Kovar and T. D. Luckey, unpublished).

or hormetic data, a definitive value can be readily determined by extrapolation for those hormetic compounds that show stimulation for the parameter being studied when minute doses are administered. Experimental design to determine ZEP requires that the complete biologic spectrum of the dose–response curve be determined. While any parameter may be measured, different criteria may not provide the same ZEP; this requires assignment of a value to the importance of each parameter. Some parameters are much more difficult to study than others. However, even the recondite toxicity studies of Schroeder (1973) revealed areas of hormology that could be used to determine ZEP for very low doses of metals administered through one to four generations of rodents. A similar concept is applicable to determine the safety of a nutritionally essential metal. The ratio between the optimally required nutritional dose and the minimum lethal dose (MLD) of the metal represents the margin of safety; the greater the ratio, the greater the toxicity of the essential metal. On this basis, selenium is highly toxic, whereas chromium is relatively less toxic.

Although the number of metals to which man and animals are exposed in an urban area has increased tremendously during the past several years, our awareness of harmful action by metals used in industry has produced remarkable safety records. Illnesses from toxic metals in Great Britain resulted in approximately 40 annual deaths at the turn of the century, and no deaths in recent years. Thus, industrial hazards from toxic metals are controllable by industrial safety and public health methods. The story is less optimistic for pollution studies. Great cities of the past were polluted from chimney smoke; today's cities have both industrial coal smoke and oil and gasoline emissions from automotive equipment. However, hotter combustion chambers produce more microparticulates than do old-fashioned stoves and heaters. Stack cleaners may give the appearance of success by removing 99% of the visible plume, and yet be useless if they do not remove a proportionate amount of the invisible plume, the particles less than 5 μm in size. As will be illustrated in the next chapter, particulates 0.1–1 μm in diameter are of special concern. Particulates less than 0.1 μm in diameter may be treated as gases for practical purposes.

The nutritional safety of the oxides of certain lanthanides was studied to evaluate their usefulness as multiple nutritional markers. Mice were fed more than 5000 ppm of eight lanthanide oxides in the diet at one time. This diet was continued for four generations. The animals exhibited no symptoms of toxicity associated with the lanthanide salts (Hutcheson et al., 1975a,b). Since no toxicity was seen, the margin of safety could not be determined. The safety of these lanthanide oxides for use as nutritional markers in the diet was established by administering 1000 times the expected use level, with no effect being noted for parameters such as general appearance, growth, maturation, fertility, reproduction, or lactation in mice. Analysis of mouse carcass and monkey tissue showed no lanthanides following two months' administration (Hutcheson et al., 1975a). These data, plus rat feasibility studies (Luckey et al., 1975), enabled us to commence studies with human subjects and explore methods to monitor food utilization by the indirect method. (Luckey et al., 1977; Hutcheson et al., 1977).

From the public health viewpoint, Mussman (1975) indicates that heavy metals in meat and poultry in the United States are not considered a hazard. Analysis of meat during the past few years indicates that beefsteak contains about 1 ppm cadmium and mercury and up to 3 ppm lead. Mercury was not detectable in one-third of the cattle tested. Despite increased fertilizer usage, the use of such metals is limited, while the increased use of metals in industry constitutes a potential hazard from wastes and accidents. This viewpoint is exemplified by beryllium. Soon after beryllium became of commercial value, reports of beryllium disease began to accumulate in Europe. By 1950, a considerable number of beryllium poisoning cases were recorded in the fluorescent lamp industry, which was then using beryllium as a fluorescent phosphor. By 1970, over 30,000 American workers were exposed to beryllium during their mining, extraction, and fabrication activities. The U.S. Atomic Energy Commission accepted a limit on beryllium of 2 $\mu g/m^3$ of air as a safe level, based on the fact that animals showed no symptoms with 10 μg, but became sick when the concentration reached 40 $\mu g/m^3$. At a meeting on beryllium contamination as an occupational hazard (Utidjian, 1973), it was concluded that the 2-μg level is universally exceeded in industry, and may reach as high as 1000 $\mu g/m^3$. Of the 837 cases in the beryllium case registry started in 1962, only 76 were entered since 1966, and over half of those since the 1969 standards were instituted. Thus, it appears that the general awareness of beryllium toxicity has not helped to decrease contamination from this highly toxic metal. A small percentage of workers exposed to chronic beryllium toxicity die of cancer during the two decades following exposure; such latent toxicity is difficult to detect and study. This finding emphasizes the importance of lifetime studies of effects of metals on experimental animals.

The general criteria for health and acute toxicity are obvious; somewhere between these two is an area where one runs into the other. This overlap poses a public health problem in the excess use of nutrients and drugs, as exemplified by the toxicity of metals. The ZEP concept designates the highest level of drug or any compound that gives results equivalent to zero (see Fig. 1-7). This level is obviously not harmful. As shown in Fig. 1-8, the mean values for any parameter may be extended by 2 σ to represent the normal population. A secondary ZEP point might therefore

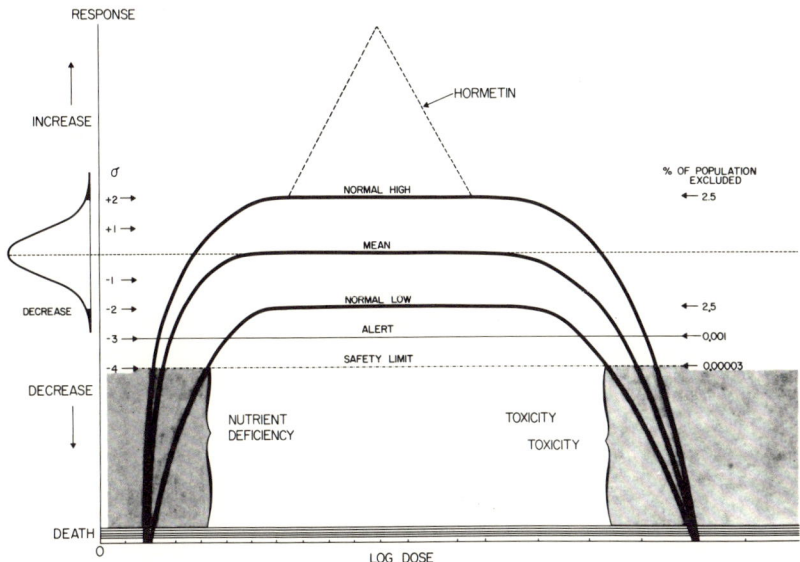

FIG. 1-8. Safety limits. The left third of the curve deals with the essential nutrients, or essential metals in this case. These curves are ideal curves, to show that different populations require different amounts of essential metal to avoid death, and that different amounts of any essential nutrient might provide a range of results in a given population. The right half is applicable to any toxicant.

be 2, 3, or 4 σ below the normal mean on a dose–response curve. Four σ below the mean represents a safe limit at which only three persons in ten million are excluded. We suggest the limit of safety be 4 σ, and that 5 σ below the mean would represent a toxic level. Therefore, the point at which the dose–response curve crossed the toxic alert level would represent an arbitrary harmful dose. Only one one-hundredth of that dose would be allowed at any one time. The actual intake from food, water, or air could be estimated and considered with the characteristics of that compound as far as its accumulation and acute and chronic toxicity data were concerned. Thus, the concept presented in Fig. 1-8 suggests a reasonable answer to the questions of how much of anything is minimally toxic and what are the limits of safety.

Since data along these lines have not been developed, this concept suggests experimental designs for the future. These designs should include a variety of usual stress factors that might be anticipated in human or domestic animal populations. Table 1-5 suggests the multifactorial stress experiment that should be considered in a dose–response curve. Data from

TABLE 1-5. Multifactoral Stress Conditions for Dose–Response Evaluations

Group	1	2	3	4	5	6	7	8	9	10
Sedentary	−	+	+	+	+	+	+	+	+	+
Sleepless	−	−	+	+	+	+	+	+	+	+
Smoke fumes	−	−	−	+	+	+	+	+	+	+
Caffeine addiction	−	−	−	−	+	+	+	+	+	+
Aspirin	−	−	−	−	−	+	+	+	+	+
Alcohol	−	−	−	−	−	−	+	+	+	+
Anxiety	−	−	−	−	−	−	−	+	+	+
Marijuana	−	−	−	−	−	−	−	−	+	+
Isolation	−	−	−	−	−	−	−	−	−	+

such toxicity studies would allow better predictions for human safety limits. Present standards for occupational exposure and health are given in the Toxic Substances List (Christensen and Luginbyhl, 1974).

ANALYTIC PROCEDURES

Toxicity due to metal compounds under acute and moderately severe chronic conditions can be distinguished by outward clinical symptoms. Postmortem examination and analysis of the tissues and internal organs indicate the extent of intoxication and the distribution of metal toxicants in the body. Toxicity due to chronic exposure to very low doses of metallic toxicants is difficult to diagnose, especially when clinical or outward symptoms are not well pronounced. Under these conditions, blood, cerebrospinal fluids, and available excretory products of metabolism such as feces, urine, skin, nails, and hair are analyzed to identify and assess the dosage of the toxicant.

Human hair serves as a special biologic sample to detect chronic toxicity caused by exposure to lead, chromium, mercury, selenium, and other toxic metals and their compounds. Sophisticated analytic techniques can be used to detect these metals, which could gain entry by contact, ingestion, or inhalation. The hair should be washed with suitable detergents and other solvents to remove the metallic contaminants adsorbed on the surface; in rare circumstances, some metallic pollutants would be irreversibly and securely bound to the cysteine of hair. By analyzing different segments of a long strand of hair, it is possible to identify some metallic toxicants and the time when the victim was exposed to the toxicant. Nails also serve as a suitable biologic sample for the same purpose.

Developments in the analysis of toxic metals during the past decade led to the identification of toxic metals as etiologic agents for diseases such as itai-itai. Routine analytic levels of detection and estimation of metals have reached picogram levels (10^{-12} g). In the beginning of the pesticide era (about 1960), the detection level was about 20 μg (2×10^{-6} g) for metals such as arsenic by the Gutseit method (Gunther and Blinn, 1955). Even the crucial initial steps in trace-metal analysis, such as sample preparation and ashing and metal isolation and concentration, have been improved considerably. Instrumentation technology has developed atomic absorption, fluorimetry, emission spectroscopy, neutron activation, gas–liquid chromatography, anodic stripping, X-ray fluorescence, electron microprobe, and spark-source mass spectroscopy techniques into versatile and highly sensitive tools for the determination of metals in biologic samples. Multimetal analysis, the ability to use milligram samples, and sensitivity at nanogram or picogram levels are the major advantages and improvements over conventional methods such as colorimetry, chemical fluorescence, and titrations. Some of the improved techniques of sample preparation and the general principles involved in the above analytic methods will be reviewed briefly.

Sample Preparation

Loss of volatile metals in dry ashing and wet digestion of samples leads to error in the analysis of metals such as chromium, arsenic, selenium, tin, and mercury.

Indiscriminate heating and dry ashing at high temperatures should be avoided, although these procedures are still good for alkaline earth metals. Acid digestion of the sample in sealed containers aids in rapid decomposition and solubilization without loss; special Teflon-lined steel vessels are used for this purpose (Paus, 1971). Oxygen flask combustion decomposes the sample rapidly without loss of volatile metals such as mercury and arsenic (Gutenmann and Lisk, 1960, 1961). Solvent extraction of metals from solid samples is appropriate when the metals are volatile at temperatures normally used for ashing samples; chromium and arsenic are typical examples. Soluene (a product of Packard Instrument Co. that is a quaternary ammonium hydroxide) is suggested as an effective solvent for extracting metal compounds from samples (Jackson *et al.*, 1972). Separation of a specific metal from a biologic sample requires true solvent extraction and chelation with a suitable compound; the sample must be solubilized in an aqueous medium to which a chelating agent dissolved in an organic immiscible solvent is added. The metal can be extracted by shaking under controlled pH conditions; e.g., cadmium can be extracted and separated at pH 10, using dithiozone in chloroform as the extractant. Solvent extraction for separation of metals has been extensively reviewed (Katz, 1972). A comprehensive survey of sample preparation for analysis of metal in trace amounts, with data on recovery and loss of various metals in different ashing procedures, temperatures, and other operating conditions, is available (Gorsuch, 1970). Other reviews describe oxygen flask combustion, oxygen stream, and low temperature activated oxygen techniques for ashing (Tolug, 1972; Mizuike, 1965).

Analytic Techniques

Atomic Absorption

Advances in instrumentation have enhanced the capability and utility of conventional flame atomic absorption spectrophotometry for the analysis of trace metals in biologic samples. The samples must be suitably digested to extract the metals, and the metals should be solubilized for eventual excitation when introduced in the flame. The basic principle involved in atomic adsorption developed by Walsh (1955) is: Following dissociation of the metal salt in the flame, the metal absorbs monochromatic light at the characteristic wavelength for the metal; this absorption is proportional to the number of atoms present. This decrease in intensity is detected, amplified, and recorded by a suitable electronics system. A hollow cathode lamp emits the characteristic spectrum of the metal in question; this light is passed directly through the flame into a dispersing monochromator, where the resonance line is isolated by a grating or prism optical system. The detector is a photomultiplier tube. Multielement lamps and double-beam alternating current systems are additional modifications to enhance the sensitivity of the method. The sensitivity of detection ranges from 0.006 $\mu g/g$ for magnesium to 72 $\mu g/g$ for praseodymium. Interference from other metals, tedious methods involved in the sample digestion, and chemical separation and preconcentration of the metal before analysis are some of the drawbacks in this method. "Flameless" atomic absorption has considerable advantages over conventional methods in sensitivity and required sample size. Preliminary reduction of mercury ion to mercury vapor and its circulation through an atomic absorption light path is an example of "flameless" atomic absorption analysis (Hatch and Ott, 1968). Induction and plasma furnaces are replacing the conventional flame in new flameless techniques capable of analyzing microliter samples.

Emission Spectroscopy

Flame spectroscopy using the emission spectra of elements has been used for many years in quantitative analysis. When a metal salt is vaporized under conditions that break the chemical bonds—such as the temperature of an electric arc—free atoms concentrate in this high-temperature gaseous environment. Some of the atomic species absorb kinetic energy and are raised to an excited state. These atomic species return to the ground state with emission of radiant energy characteristic of that species. The intensity of the emission at a specific temperature is proportional to the number of atoms present in the vaporizing flame; this relationship is the basic principle involved in flame emission spectroscopy for quantitative analysis. Sophisticated instrumentation measures this emission intensity and wavelength; the light emitted in the excitation source is dispersed in a spectrometer, which images the atomic lines of the element on the focal curve. Photomultiplier detectors are used to intercept the elemental lines; the currents produced by the detectors are proportional to the concentrations of specific elements in the sample. These current impulses are digitized and converted into concentrations by means of comparison standards.

Recent advances in instrumentation have resulted in the effective use of emission spectroscopy for multimetal analysis of biologic material (Cowgill, 1972). Optimum excitation temperature and minimum background excitation, which are essential prerequisites in emission spectroscopy, are attained by the use of "argon–silver" excitation of silver chloride vapor in an argon atmosphere. This argon plasma source, commercially available as "spectra jet," can detect sodium, potassium, iron, silicon, calcium, strontium, magnesium, copper, zinc, chromium, nickel, manganese, aluminum, and vanadium in ashed serum; the detection levels range from 0.05 to 35 ng/mg of ashed serum. Another innovation in instrumentation is low power inductively coupled with the microwave discharge technique: a glow capillary is inserted in the excitation cavity, and samples are introduced by drying a few microliters of solution on a platinum-wire filament. The small sample is vaporized by applying current to the filament, and a flow of argon sweeps the sample to the excitation zone. The excitation temperature of the plasma formed inside the capillary is in excess of 5000°K. Sensitivity for detection of elements, of the order of

10^{-11} to 10^{-12} g, is due to the efficient introduction of sample into the high-temperature plasma.

Chemical interference affects both emission and absorption spectroscopy methods.

X-Ray Fluorescence

X-ray fluorescence analysis permits rapid and simultaneous analysis of a large number of metals over a wide range of concentrations; the sensitivity is usually at the microgram level. Liquid samples are lyophilized; other samples are dried by low-temperature ashing. Samples may need preconcentration by suitable extraction, followed by filtration through ion-exchange resin paper disks, which are then used directly in the analysis.

Fluorescence or secondary X rays are generated when an atom is bombarded with an energetic electron, photon, or primary X ray that removes an interior electron from the metal. These ionized unstable atoms return to the ground state by transfer or transition of electrons from an outer shell to the incomplete inner shell, with a concomitant emission of fluorescence X rays equal in energy to the difference in electron energy levels. The energy of this fluorescence X-ray emission is a function of the atomic number of the element and is characteristic for that element. This relationship serves as the basis for the quantification of metal by X-ray fluorescence. The older wavelength dispersive analysis and the newer nondispersive X-ray energy spectroscopy are the two techniques employed to measure X-ray fluorescence (Woldseth, 1973; Frankel and Aitken, 1970).

In wavelength dispersive analysis, proportional counters and scintillation counters are used. The electron beam strikes an anode composed of a thin foil of the target material inside transmission high-energy X-ray tubes; solid powder samples and special filter disks, through which sample fluid can be filtered, can serve as anodes. X rays generated within the foil are observed from the opposite side following their passage through the anode; they are detected by gas-flow proportional counters of sodium iodide–thallium scintillation crystals. Energy discrimination of X rays is done by a combination of crystal dispersion and pulse-height selection. Sophisticated electronic instrumentation aids in resolving the X rays when more than one metal is involved. In modern equipment, a multichannel analyzer collects the digitized signals from the detector and sorts the signals according to pulse heights. Interference from other metals is compensated by use of standards that contain the interfering metals. The sensitivity of this method is a complex function of many variables, including sample thickness, primary energy and intensity, X-ray tube target, and other interfering metals.

In X-ray energy spectroscopy, radioisotopes such as ^{241}Am, ^{125}I, ^{57}Co, or ^{153}Gd serve as the photon (X-ray excitation) source; the spectrometer uses a high-resolution silicon or germanium solid-state detector. The output of the detector is amplified and sent through a pulse-height analyzer system. Computer-aided data reduction is usual on such nondispersive-type instruments. The requirements for sample preparation are minimal, and the system accommodates samples of a variety of sizes, shapes, and forms; this X-ray analysis is nondestructive, and a wide range of elements can be analyzed simultaneously; the concentration ranges from 100% concentration to less that 1 μg/g. The excitation source is a compact and simple radioactive fuel.

The limitations of this technique include microgram sensitivity and poor detection of elements of atomic number 20 and below.

Electron Microprobe

The electron microprobe was first designed in 1951 (Castaing, 1951), but only during the past decade has the microprobe technique been used to detect metals in biologic samples, especially element analysis of sample surfaces (Mellors and Carrol, 1961; Brookes et al., 1962). The theory and practice of electron-microprobe analysis was reviewed by Anderson (1967). Robison (1970) reviewed the literature on the biologic applications of electron-microprobe analysis and the sample-preparation techniques for this analysis.

Electron-microprobe analysis is based on the characteristic X-ray production following bombardment of a sample with an accelerated electron beam. The characteristic X rays of a particular element are selected for analysis by crystal dispersion. Proportional counters are used for the detection of X rays, and these detectors are combined with pulse-height selection for further energy discrimination. The detection limits for many elements in biological tissue are in the range 100–800 ppm. Since the tissue analyzed measures only a few cubic microns in volume, the absolute amounts detected range from 10^{-12} to 10^{-17} g. The composition of a single cell and the distribution of certain metals within the different regions of a cell or tissue can be analyzed.

In this analysis, the electrons produced at a filament are accelerated and passed through a magnetic condenser lens, which controls the diameter of the electron beam. The 2-μm beam elicits the emission of characteristic X rays from a biologic tissue, resulting in efficient resolution to determine the elemental distribution in cells. The accelerated electron beam is used to bombard the sample in two ways: static mode and scanning mode. In the static mode, the electron beam hits one localized area of the sample, and the characteristic X rays associated with a particular metal or element are analyzed by crystal dispersion. In the scanning mode, the electron beam is swept across the sample and is synchronized with the scan of an oscilloscope. The X rays are picked up by scintillation or solid-state detectors; the impulses are amplified and combined with pulse-height selectors, and finally are read in a suitable scaler. Prior calibration with suitable standards enables quantification of metals in samples. Although the probe is meant primarily for the analysis of elements on the surface of a sample, this technique is also useful for the quantification of elements in a given sample. Other probes, which are used to characterize and analyze sample surfaces *in situ,* are the laser beam (Eick *et al.,* 1967; Treytl *et al.,* 1972) and ion beam probes (Karasek, 1970; Chandler, 1971); the detection sensitivity of these probes is 10^{-15} and 10^{-19} g, respectively.

Sample preparation is an important and sometimes limiting factor in microprobe analysis of tissues. The sample must be dehydrated, since the microprobe operates at a vacuum of 10^{-5} Torr, and coated with a conductive material such as carbon, aluminum, or gold. The sample must be mounted on a suitable disc made of either quartz or a Formva-coated grid. Good contact between the specimen and the support is necessary to provide a good heat sink and to avoid heat destruction of the sample. A uniform composition throughout the thickness of the sample preparation is absolutely necessary to obtain reliable quantitative values.

At very high acceleration potentials, the electrons interact very little and lose only a small fraction of their initial energy as they speed through the thin specimen. The intensity of the continuing spectrum produced as a result of the deceleration of the electron beam in the sample is used to measure the mass per unit area in the sample. The continuing X-ray intensity and the characteristic line X-ray intensity are used to quantify the element under study in the sample (Hall, 1968).

Gas-Liquid Chromatography

Gas-liquid chromatography (GLC) has been used to detect metals such as cobalt, selenium, chromium, and beryllium in trace amounts present in biologic materials (Savory *et al.,* 1969; Ross *et al.,* 1970; Taylor *et al.,* 1968). Moshier and Sievers (1965) give a good review of the GLC of metals. Some metal chlorides such as titanium chloride can be volatilized and analyzed by GLC, but the rest of the metals should be converted into stable, volatile, organically bound compounds. These volatile organic chelate compounds are chromatographed and detected by electron-capture techniques. The sensitivity and analytic limit of detections ranges from nanogram to picogram levels.

The organic compounds used in the GLC of metals are mostly fluorine-substituted β-diketones such as trifluoroacetylacetone (TFA), hexafluoroacetylacetone (HFA), heptafluorodimethyloctanedione (FOD), hexafluoromonothioacetyl acetone (HFTA), and decafluoro-4,6-heptanedione (DFHD).

$$F_3C-\underset{\underset{O}{\|}}{C}-CH_2-\underset{\underset{O}{\|}}{C}-CH_3 \qquad F_3C-\underset{\underset{O}{\|}}{C}-CH_2-\underset{\underset{O}{\|}}{C}-CF_3$$

TFA   HFA

$$H_3C-\underset{\underset{CH_3}{|}}{\overset{\overset{CH_3}{|}}{C}}-\underset{\underset{}{}}{\overset{\overset{O}{\|}}{C}}-CH_2-\overset{\overset{O}{\|}}{C}-\underset{\underset{F}{|}}{\overset{\overset{F}{|}}{C}}-\underset{\underset{F}{|}}{\overset{\overset{F}{|}}{C}}-\underset{\underset{F}{|}}{\overset{\overset{F}{|}}{C}}-F$$

FOD

The metal chelates of these highly fluorinated chelating agents contain a large number of electronegative atoms, rendering them highly sensitive for the electron-capture detector of the GLC system. Metals that do not react with these compounds are molybdenum, technetium, silver, praseodymium, selenium, tungsten, and metals beyond tungsten in the periodic table. Lithium, divalent copper, and tetravalent vanadium form fluorine-substituted β-diketone derivatives, but the chelates cannot be chromatographed. Selenium is converted into a stable, volatile, and toluene-soluble derivative, but selenium in biologic material must be separated and converted into selenous acid for reaction with phenylenediamine at pH 1–0 (Shimoishi, 1973). HFTA is a sensitive reagent for divalent metals such as copper, zinc, nickel, cadmium, and lead. The halogen-substituted β-diketones react directly with metals and metal compounds in the biologic samples; therefore, the tedious digestion procedures necessary in the other methods to convert the metal in the sample to its ionic form can be avoided. The factors that influence the volatility and stability of the metal fluoroacetyl acetonates have been reviewed and discussed (Sievers, 1969). The procedures described above are so specific that meticulous attention should be given to the operating parameters in GLC of these metal chelates.

Sophisticated instrumentation has been used to detect and estimate organometallic compounds other than the metal fluoroacetyl acetone chelates by GLC. Organometallic compounds of iron, lead, and tin were detected at 10- to 30-pg levels with a hydrogen-rich flame ionization detector; a flame photometric detector is used to determine organometallic compounds of nickel, mercury, lead, and copper in gas-chromatographic effluents (Aue and Hill, 1972).

Gas–liquid chromatography of metal fluoroacetyl acetone chelates is the best and most reliable method for estimating metals present in trace amounts in biologic material.

Spark-Source Mass Spectrometry

Spark-source mass spectrometry is useful in multielement analysis on a prepared and ashed sample. Ions produced by high-voltage excitation of the metal are energy-focused in an electrostatic analyzer to pass through a magnetic analyzer before impinging on a photographic plate. The ions hit the photographic plate at different points, depending on their mass-to-charge ratio. Analysis is generally carried out by referring the intensity of the desired metal or element line to a series of known standards. Recent and sophisticated electronic detection systems have enhanced the analytic limit; these include electrostatic peak switching with static integration and computer interfacing of the electrically recorded spectra (Bingham and Elliott, 1971; Brown *et al.*, 1971).

The analytic level of detection is in the range 10^{-6}–10^{-11} g, and this technique has been used for biologic and forensic samples. High sensitivity, wide element coverage, and rapid analysis are the advantages of this technique, but the sample must be processed and dry ashed without losing the volatile metals. Spark-source mass spectrometry was reviewed by Ahearn (1972).

Anodic Stripping Voltametry

Anodic stripping voltametry is extremely sensitive and useful for the detection of metals in solution, especially metals in natural aquatic systems; the detection limit is in the nanogram (10^{-9} g) range. Sensitivity to many metals and the capability of multimetal analysis in the same sample without any chemical separation are the two advantages of this technique; dry and solid samples must be digested suitably to obtain the metals in solution. Application of this technique to tissue, blood, urine, and hair has been described (Matson *et al.*, 1970).

Neutron Activation

Advances in gamma-ray detectors, pulse-height analyzers, neutron sources, and data reduction have made neutron activation analysis, first introduced by Hevesy and Levi in 1936, an efficient technique in trace

INTRODUCTION TO HEAVY METAL TOXICITY IN MAMMALS 35

metal analysis. Following irradiation with neutrons, a metal in a sample becomes radioactive and emits a characteristic radiation which can be measured. The constant energy and half-life of a gamma ray from the irradiated sample identify the radionuclide and serve to quantitate the metal in a given sample. The dry, unashed sample is irradiated in a nuclear reactor for suitable periods (one minute to several days in a high neutron flux). Gamma-ray spectra are obtained from decay curves, which may range from minutes to several days following radiation and depend on the metal assayed. Gamma-ray photo peaks in the pulse-height spectra are first identified and then quantitatively determined by means of computer-aided data reduction. The sensitivity of this method is high, and the levels of detection vary for different elements: microgram levels for iron, silicon, and selenium; nanogram levels for a host of metals (e.g., sodium, aluminum, arsenic, zinc, cadmium, and gold); and picogram levels for europium, dysprosium, indium, and manganese. The higher the neutron flux, the greater is the ability to determine a metal. This technique cannot be used for metals such as lead, lithium, beryllium, thallium, osmium, and bismuth, because of their nuclear properties; other, more convenient methods are available for determining these metals.

The sensitivity of neutron activation analysis for a given element is influenced by a number of factors: (1) the abundance of the stable isotope that can be irradiated; (2) cross section of the atom of the isotope for neutron capture; (3) neutron flux; (4) half-life of the radionuclide produced; (5) decay scheme of the radionuclide; (6) decay period before counting; (7) efficiency of the radiation detector, such as thallium-activated sodium iodide crystal; (8) irradiation time; and (9) interference from other metals. With improved instrumentation in the detector, high neutron flux, and varied irradiation times, it is possible to determine trace amounts of metals quantitatively. The thallium-activated sodium iodide detector is efficient for single-metal detection and estimation. Lithium-treated germanium solid-state gamma-ray spectrometers are used in multielement analysis because their resolution is greater than the iodide detectors, although they are only about 10% as efficient as the sodium iodide (thallium) detectors. A low-energy photon detector has increased efficiency and resolving ability in the X-ray region and can considerably extend the capabilities of a conventional system.

Neutron activation analysis is superior to other methods in sensitivity for multielement analysis, provided the sample is homogeneous and that standards are prepared to resemble the samples as closely as possible. Internal standards should include the sample matrix and the interfering elements. Large amounts of sodium, chloride, and phosphate ions present in the sample may have to be removed before counting; the analytic limits of many other metals are considerably enhanced if these elements are not found in multielement samples. Samples containing sodium, potassium, and chloride may be allowed to decay until these relatively short-lived radionuclides are gone.

Principles, experimental procedures, and applications of neutron activation analysis for multimetal determination in biologic samples have been extensively reviewed (Kay et al., 1973; Kruger, 1971; Kramer and Wahl, 1968; I.A.E.A., 1972; Hoste et al., 1971).

Summary of Techniques

The analytic limits of a few metals using these sensitive analytical techniques are summarized in Table 1-6. Spark-source mass spectrometry

TABLE 1-6. Analytic Limits for Metals in Various Techniques[a]

Metal	GLC	Atomic absorption	Neutron activation	Emission spectroscopy	Spark-source mass spectrometry
Be	6×10^{-14}	1×10^{-8}		2×10^{-10}	8×10^{-12}
Cd		5×10^{-9}	1×10^{-10}	2×10^{-6}	
Hg		5×10^{-7}	1×10^{-8}	2×10^{-8}	
Ga		7×10^{-8}	1×10^{-10}	1×10^{-8}	
In		5×10^{-8}	3×10^{-12}	2×10^{-9}	
Tl		2×10^{-7}	—	2×10^{-8}	
Sc		1×10^{-7}	1×10^{-9}	3×10^{-8}	
Ge		1×10^{-6}	3×10^{-9}	5×10^{-7}	
Pb		3×10^{-8}	—	2×10^{-7}	
Ti		1×10^{-7}	1×10^{-9}	2×10^{-7}	
Zr		5×10^{-6}	5×10^{-7}	3×10^{-6}	
As		2×10^{-7}	2×10^{-10}	—	
Sb		1×10^{-7}	2×10^{-10}	—	
V		2×10^{-8}	1×10^{-11}	1×10^{-8}	
Nb		3×10^{-6}	—	1×10^{-6}	
Se	4×10^{-9}	5×10^{-6}	2×10^{-9}	—	1×10^{-6}
Te		3×10^{-7}	2×10^{-9}	—	
Cr	2×10^{-14}	2×10^{-9}	1×10^{-6}	1×10^{-9}	5×10^{-11}
Mo		3×10^{-8}	1×10^{-9}	1×10^{-7}	
W		3×10^{-6}	2×10^{-10}	5×10^{-7}	
Mn		2×10^{-9}	4×10^{-12}	5×10^{-9}	
Fe		5×10^{-9}	1×10^{-5}	5×10^{-8}	
Co	1×10^{-11}	2×10^{-9}	5×10^{-10}	5×10^{-8}	5×10^{-11}
Ni		5×10^{-9}	2×10^{-8}	3×10^{-8}	
Pd		2×10^{-8}	2×10^{-9}	5×10^{-8}	
Pt		1×10^{-7}	1×10^{-8}	2×10^{-6}	

[a] Limits expressed in g metal/g sample.

and electron microprobe analysis appear to be the most sensitive in the detection levels. GLC and neutron activation analysis are next best. The novice should be alerted to the fact that the sensitivity of detection may be 1000-fold greater than the sensitivity for practical analytic procedures in biologic material.

Review of the development and instrumentation improvements of analytical techniques for metals suggests that more sensitive instrumental techniques will be developed. The limiting factor will be background contamination from reagent impurities, glassware and air pollution in the laboratory, especially as the sample size continues to become smaller and smaller. Laboratories near highways and fossil-fueled power-generating stations will certainly have more quantities of lead, cadmium, barium, mercury, and thallium in their atmospheres than laboratories far away from these contamination sources. Oil-fired power plants have vanadium and other metals. Contamination or loss of metals during high-temperature ashing or processing of samples will be another critical factor. Plastic laboratory ware serves as a contamination source, since the metals used as catalysts in the synthesis of plastics may be retained in trace amounts.

2

MODES OF INTAKE AND ABSORPTION

Metals and their compounds are usually ingested or inhaled. Industrial accidents (slivers, cuts, and abrasions) cause subcutaneous exposure in man. The route of intake and the dosage determine the intensity and duration of the harmful effects (the dosage–time curve) of a toxicant. Oral administration provides processing of the toxicant by digestive enzymes, microbic cells, host intestinal cells, and often hepatic cells before the material is transported to other tissues (Fig. 2-1). Inhalation results in the direct transfer of soluble metal compounds from the pulmonary tissues to the blood; it can also provide a circuitous route for the transport of the toxicant to the tissues. Cilia of epithelial cells can carry insoluble metal particulates or compounds from the lungs into the gastrointestinal tract via the pharynx. Macrophages can transfer the metal directly to regional lymph nodes, blood, and other tissues or to the gastrointestinal tract. Dermal administration usually provides prolonged absorption of small quantities of material, especially lipid-soluble compounds, over an extended period. Intravenous (including intraarterial) administration distributes the dose to tissues within seconds, and—other than absorption to plasma proteins, cell walls, or other membranes—there is little systemic processing of the material. Metabolic processing is generally less important for minerals than for organic toxicants.

The release of metal ions to tissues determines the dose–response sequence, with time as an important factor. Intravenous administration quickly delivers a high initial dosage to tissues, in contrast with the gradual exposure of tissues following ingestion, inhalation, or dermal absorption. Other injection routes fall between these two extremes. The importance of time in the dose–response relationship is the substantial delay of most compounds in reaching a maximum concentration within the milieu of the

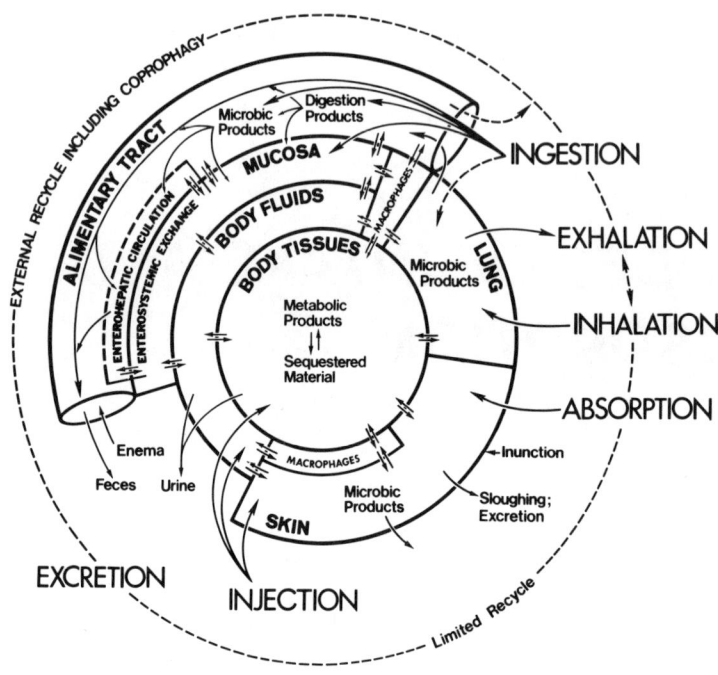

FIG. 2-1. Modes of intake and excretion for metals in mammals.

metabolizing cells. If a dose–response reaction of a nonnutrient approximates the β-curve response (see Fig. 1-4), swallowing a harmful dose of the compound might elicit a reaction sequence in which stimulation precedes toxicity. This sequence would occur if absorption of the compound were slow enough to permit a reaction to the lower concentrations before the full impact of the dose was exerted. As high concentrations are absorbed, a toxic reaction is registered. Thus, the time sequence of observed reactions may be dramatically different, depending on whether a given amount of compound reaches tissues relatively slowly following oral administration, or quickly following injection. Evaluation of toxicity might depend primarily on the reaction at the time selected for observation. The time required for the onset of toxicity symptoms following intake is important in reading LD values of toxicants, particularly in cumulative studies.

Obvious differences in responses to toxicants following different routes of administration relate to how the material is absorbed, metabolized, sequestered, and excreted. Oral administration evokes a series of protective mechanisms: a physiologic response, such as vomiting; metal interactions; chemical action of hydrochloric acid in the stomach; enzyme action in the small intestine; and metabolism by intestinal microbes (10^{14} in man). Inhalation processes may be no less complex, since particulates may be carried by macrophages from lung tissue to other tissues, including the alimentary tract (Fig. 2-1).

The species, time of day, and condition of the organism also modify the absorption, homeostasis, and toxicity. A variety of diseases alters the amount of toxicants absorbed and the response of the organism to a given dosage. Age, sex, body weight, body composition, and stress will greatly change the absorption, sensitivity, and reaction to toxicants. Other important variables are the state of nutrition, the season, and the time of day. Chronotoxicology was reviewed by Halberg (1969). Physical or psychological trauma associated with parenteral injection or enema may cause infection or other responses that influence the absorption of the toxicants and the eventual toxicity symptoms. This is especially significant in humans. Unless otherwise stated, healthy subjects are assumed in this discussion.

In order to exert their action, toxic substances must either affect or penetrate cell membranes. Absorption includes those processes by which a compound enters the blood or tissues. At the cellular level, the material must cross several membranes for intake, utilization, homeostasis, storage, or excretion. This passage is usually a molecular transport by the process of diffusion, facilitated transport, or active transport. The lipoidal nature of cell and organelle membranes has much influence on these processes. Endocytosis, the process by which the cell engulfs liquids (pinocytosis) or solids (phagocytosis), is important in absorption. Pinocytosis is more important in the newborn than in the adult intestine (Williams and Beck, 1969). Persorption is discussed as a minor component of intestinal absorption. Persorption, trauma, and infection allow direct access to tissues. This brief summary of absorption was taken from standard texts (Lehninger, 1975; White *et al.*, 1973). McColl and Sladen (1975) review the absorption of sodium, potassium, calcium, and iron.

Most molecules transport themselves across membranes by simple diffusion, or passive transport, which is the simple passage of soluble material across membranes at rates and quantities determined by the physical character of the system, a dynamic system in living cells. Membrane permeability and electrochemical and concentration gradients are most important for diffusion of material into cells. Different toxicants may alter organelle or plasma membranes by changes in membrane permeability, electrochemical constituents, or osmotic pressure of the cell, which quickly alter the influx of compounds entering by passive diffusion. Water passes through membranes by diffusion. Nonpolar and lipid-soluble materials cross membranes readily. Nonionized compounds cross membranes more rapidly than do metal ions. However, the aqueous perimatrix filling the area between the intestinal microvilli blocks much lipid material from the absorptive surface. Material that can penetrate this mucus is readily absorbed by micropinocytosis.

Passive facilitated transport accelerates the absorption of specific compounds by facilitated diffusion. Specific carrier proteins of the membrane react with specific compounds and their close analogs for escort through the membrane; these proteins resemble the permeases of bacteria. The direction of the substrate depends on the concentration gradient of the compound on either side of the membrane. Analogs competing with the substrate show kinetics comparable to those of enzyme inhibition. Cells provide mediated transport to certain vital nutrients or metabolites that are needed at rates faster than those provided by simple diffusion, and the selectivity provides increased efficiency. Examples include Ca^{2+} binding to a vitamin-D-derived carrier protein and Fe^{2+} binding to apoferritin during intestinal absorption. Cobalt and probably manganese reduce iron absorption in the intestinal mucosa by competition for a common carrier (Schade *et al.*, 1970).

Active transport is a specialized facilitated transport that requires energy to transfer a compound across a membrane against a concentration gradient. The membrane contains a specific carrier that will transport a specific compound in a single direction according to the cell orientation. In the absence of an energy source, active transport systems provide passive facilitated diffusion in either direction. Active transport typifies plasma membranes; it helps to provide nutrients and maintains the selective array of molecules for intracellular homeostasis in a great variety of environments. Active transport usually involves the energy of hydrolysis of ATP, and often involves the exchange of intracellular Na^+ for the material (sugar or K^+) being brought into the cell. The substrate is sometimes cotransported with Na^+; the Na^+ is then excreted by an ATPase. Energy may be expended for metabolic alterations of the molecule as the compound is released from the carrier. Sodium, calcium, and iron have active transport mechanisms. The sodium pump is at the basal membrane of the mucosal cell. The calcium carrier protein functions at the brush border. Iron has active transport at both surfaces, with an effective control system within the mucosal cell. Chromium is most readily absorbed in an organometallic form, the glucose tolerance factor. Cobalt can be absorbed best in the form of cobalamin. The solubility of the metal ion or complex in the intestinal lumen is the key to effective absorption in mammals. Low-molecular-weight organic acids increase metal absorption by increasing the solubility of metals other than those of Group IA.

Endocytosis requires energy for the movement of material and the plasma membrane. Endocytosis and persorption are nonspecific; a mixture of compounds and particulates may be engulfed by these processes.

ENTERAL ADMINISTRATION

Oral Intake and Absorption

Ingestion has the greatest potential for selective acceptance or rejection of all modes of administration. Physical and chemical attributes of ingesta provide sensual attraction or repulsion. Vomiting is a physiologic reflex permitting considerable selectivity of toxic agents; mammals that rarely vomit are susceptible to materials that cause unusual stomach reactions. Finally, materials that irritate the alimentary tract are purgatives. The effect of toxicants is greatly affected by the presence of food in the stomach. Eating a toxicant before a meal, or early or late in a meal, will affect the rate at which the toxicant is released into the intestine.

A second process of oral intake is drinking; liquids usually pass quickly through the empty stomach into the slightly alkaline intestine (pH 6.5–7.8). Drinking and gavage are little used in mammalian toxicology.

Soluble metal compounds are absorbed throughout the alimentary tract, with the amount of absorption varying. For limited quantities held in the mouth, sublingual absorption is fast and effective; it allows contact only with salivary diastases and some microbial systems. Drugs are absorbed faster here than from the gastrointestinal tract. Limited absorption by oropharyngeal and esophageal mucosa and their associated bacteria occurs during the few seconds of swallowing (Savage *et al.*, 1968). Most absorption of metals occurs in the duodenum and jejunum.

Gastrointestinal Absorption and Persorption

Introduction

As illustrated in the diagram (Fig. 2-1), digestion within the alimentary tract provides an extracorporeal processing by stomach acid, digestive enzymes, and microbic metabolism. Products of digestion may be eliminated or absorbed, according to their effects on the intestinal mucosa (Davenport, 1961).

The intestinal mucosa is a large, active metabolic tissue. Its area (about 200 m²) is comparable to that of a tennis court (Creamer, 1974), and its rate of oxygen uptake equals that of liver tissue. Epithelial cells form a single layer to define the pulsating villus (Fig. 2-2). The intestinal mucosa is capable of rapid regeneration; newly formed cells migrate along the side of the villus to be extruded at the villus tip (Fig. 2-3D). Most cell division occurs in the lower part of the Lieberkuhn crypts. Some differentiate into goblet cells, others into specialized cells with a full complement of digestive enzymes during the several hours required for their emergence from the crypts. As they migrate to the tip of the villus, they differentiate into absorptive cells with microvilli and microcalyces, the microcalyx being an external extension of the microvillus. They must become physically disoriented at the tip prior to extrusion into the lumen. This continuous extrusion of cells becomes a defense mechanism, removing toxic doses of metals bound to cellular macromolecules. Small quantities of highly toxic materials would be expected to kill the first cells with which they came into intimate contact. Death of a small number of columnar epithelial cells is not serious because they are rapidly regenerated; they all are replaced within

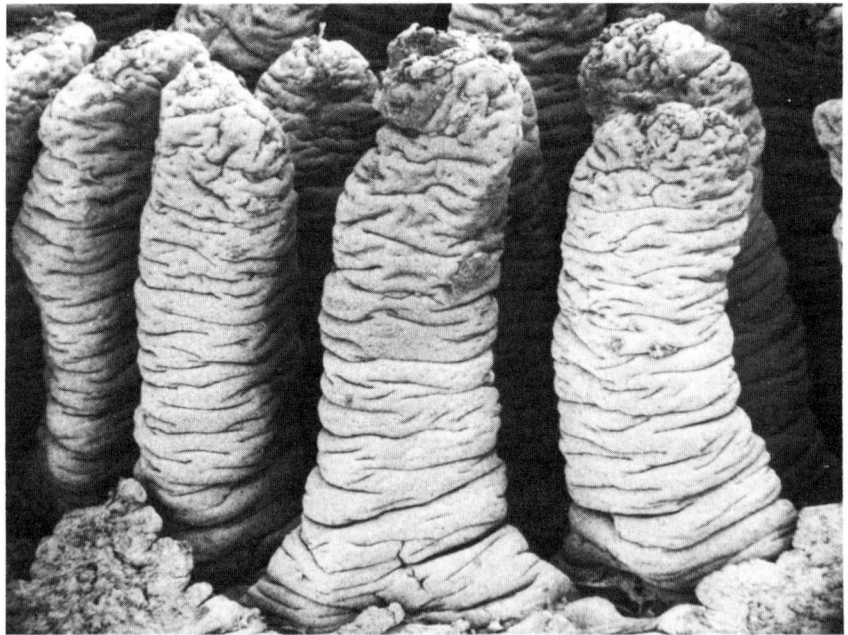

FIG. 2-2. Villi of dog jejunum (field width ~ 850 μm). (Scanning electron microscope micrograph contributed by Dr. Janice Nowell, Thimann Laboratories, University of California, Santa Cruz.)

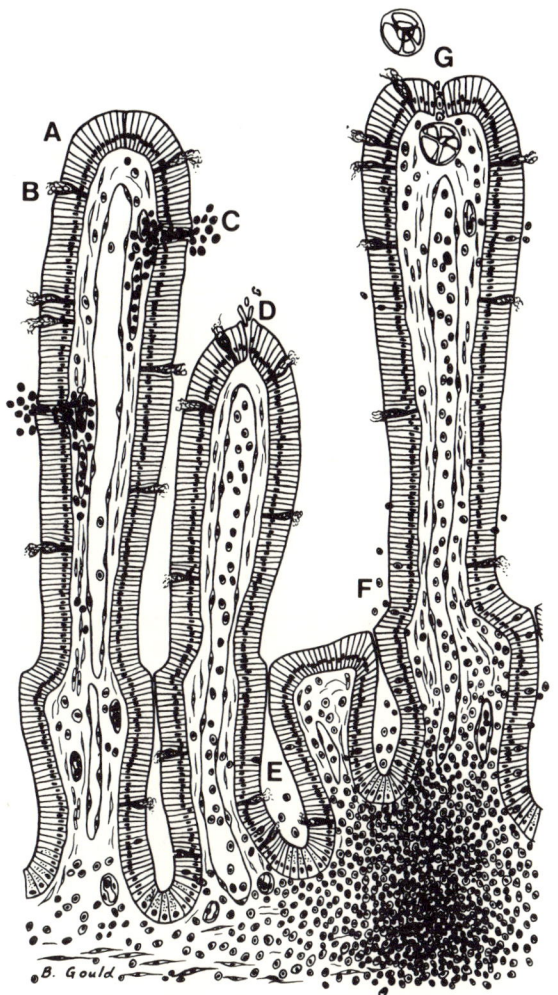

FIG. 2-3. Contributions of the intestinal villus to the succus entericus: A) Secretions from absorptive cells of the mucosa; B) goblet cell mucus; C) neutrophil excretion; D) sloughed mucosal cells; E) Lieberkuhn crypt secretions; F) lymphocytes; G) Starch granule implosion, persorption, and leakage.

2–3 days. The cells most exposed to digesta are those closest to the tip of the villus, where dying cells are cast into the lumen (Fig. 2-2D). This sloughing of old epithelial cells (10^{10} cells/day in man) permits the body to be rid of excess metals or their salts and is a good defense against some toxic minerals that can be sequestered in the epithelial cells. Another physiologic protection is the removal of irritating toxic material by purgation or diarrhea. The cathartic action of many inorganic metal salts is well known. Finally, the passage of portal blood from the intestine directly to the liver provides several mechanisms for phagocytosis, detoxication, or sequestering of harmful compounds. Metal toxicity may be expressed at the absorptive surface of a cell to prevent nutrient absorption.

The Villus

The role of the villus in digestion and absorption is reviewed by Luckey (1974). The villus is a miniorgan complete with epithelium, stroma, muscles, nerves, blood, lymphocytes, and lacteals. Man has 5×10^6 villi, about 30 per square millimeter of mucosa. Fingerlike villi, which typify the upper intestine in Caucasians, have dimensions of about 100×800 μm. The absorptive cell of the villus epithelium is diagrammed in Figure 2-4. The length of the columnar epithelium in healthy intestinal villi is about 8 times greater than its width. The brush border presents rigid microvilli with a mucus perimatrix, the glycocalyx, firmly associated with it. The half-life of this perimatrix is only 3 hr.

Succus entericus is a mixture of crypt secretions, mucus from goblet cells, cell extrusions from microvilli, dying epithelial cells, neutrophils, and lymphocytes. The 3 liters of succus entericus secreted each day divided by 5×10^6 villi indicate that each villus supplies about 2 μl of fluid per day. Much of this comes from secreting crypt cells, while the mucus comes from the goblet cells and the microvilli. The total export from the mucosa includes neutrophils, which are exported into the lumen, particularly following secondary antigen stimulation. Macrophages migrate through the entire villus. The fluid exchange between villus and absorbed digesta provides continuous replenishment of media and removal of digestion products (Fig. 2-5), in addition to the chyme spread over each ever-active villus as it flows slowly through the lumen. The absorption of almost 8 liters of fluids is accompanied by pistonlike contractions that expel lacteal contents into lymph vessels and contribute to the cell extrusion from the tip. As the villus relaxes and elongates, the decrease in internal pressure encourages micropinocytosis and persorption.

None of the dying epithelial cells, and probably few leukocytes, survive in the intestine. Although their enzymes are of little importance in digestion, they provide a full complement of nutrients to the intestinal microflora. Textbooks indicate that one man may lose 200 g of mucosal cells having 40–50 g of protein each day. On a weight basis, each of us extrudes villi cells equivalent to our body weight each year. On a cell basis, an average villus may replace 100 cells in height and a circumference of 90

MODES OF INTAKE AND ABSORPTION

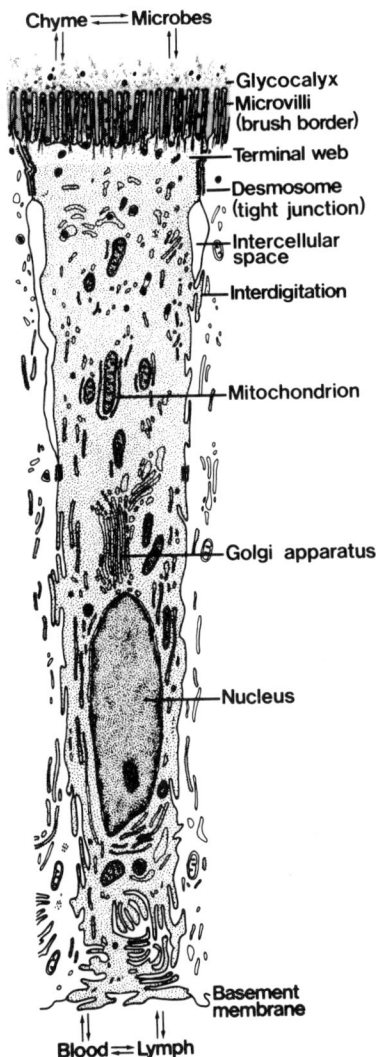

FIG. 2-4. Absorptive cell of the intestinal mucosa.

cells within 1.5 days; 9×10^3 cells multiplied by 5×10^6 villi suggests that approximately 4×10^{10} cells are lost from each person each day. The alkaline phosphatase, peptidases, and disaccharidases of the 3000 microvilli of each cell are important components of digestion (Crane, 1975). The secreted perimatrix of the microvilli (Fig. 2-6) provides a glycoprotein mucus filter for absorbed material, a buffer against particulates, and a matrix for enzyme action and microbic association; water, hydrolysis

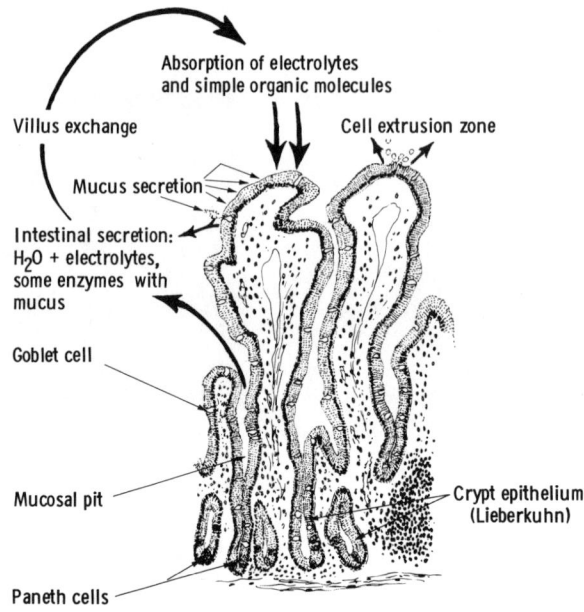

FIG. 2-5. Fluid exchange in the villus microcirculation cycle.

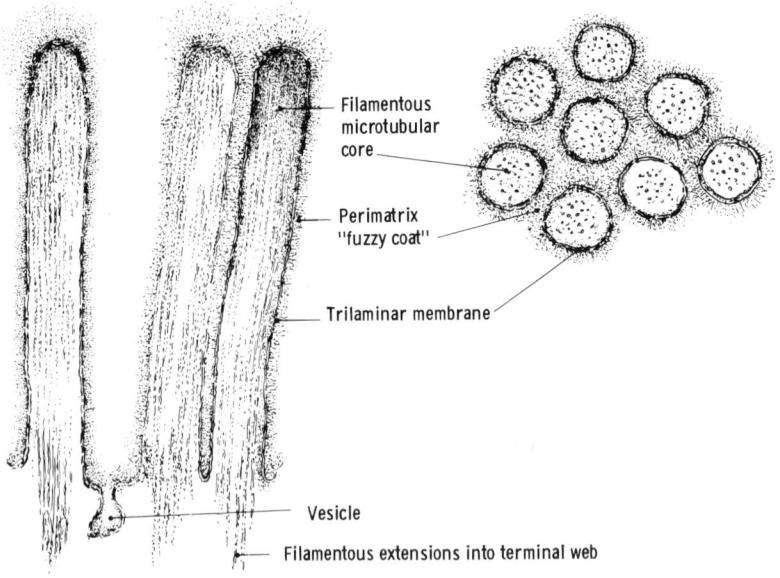

FIG. 2-6. Microvillus structure showing the glycocalyx, or perimatrix. Redrawn from Toner et al. (1971).

products of glycerides, and water-soluble small molecules go readily to the base, where micropinocytosis is a major uptake mechanism.

The flow of chyme as it slowly oozes around villi was estimated by Luckey (1974). A model of a single bite of food taken midway through lunch (Fig. 2-7) indicates that a good proportion is in the alimentary tract for 4–5 days unless a high-fiber diet is eaten. At each peristaltic contraction of the stomach (about 3/min), the pyloric sphincter ejects 2–3 ml of chyme at pH 5–6; thus, the rate at which material enters the small intestine approximates the values given in Fig. 2-8. The duodenal flow is continuous for several hours each day; distally, the spread of digesta and the absorption of fluid provide a continuous flow for about 15 hr each day. The mixing, churning, and peristalsis decrease in frequency and vigor, and antiperistalsis increases as chyme progresses down the 3 m of small intestine. At the terminus, the ileocecal valve controls the flow of chyme into the colon. The valve has two components: a sphincter and an end that prolapses into the cecum. A full cecum and ascending colon or an irritated appendix stops flow through the ileocecal valve; while eating, stomach and upper intestinal peristalses and/or a full ileum stimulates small spurts of material through the valve. Consequently, characteristics of a batch chemostat in the lower ileum coalesce with the continuous flow chemostat of the upper ileum to provide the equivalent of a cecal fermentation. When the colon is full, no appreciable back-flow is noted in radiographic studies; when the colon empties, back-flow into the ileum is noted; exact information for normal passage is becoming available (Clemmens *et al.*, 1975; Luckey, 1977). The cecum of some animals empties and fills in slow (diurnal?) cycles.

This slow progression of material continually churned by peristaltic plicae movement and villi pumping, combined with the secretion and absorption of several liters per day of succus entericus, provides ideal conditions for the growth of villi-associated microbes. The digesta accumulated in the lower end of the ileum, cecum, and colon contains 10^{11}–10^{12} anaerobic bacteria per gram.

The metabolism of microorganisms of the intestinal tract produces more complex reactions. The intestinal microflora, flourishing in symbiosis, compete with the host intestinal mucosa for essential minerals necessary for microbic reproduction. These microbes synthesize chelating compounds to solubilize the metals in the digesta for eventual absorption into their own cells. The slow flow of digesta allows ample time for intestinal microbes to process and utilize the minerals (Luckey, 1974). Weinberg (1974) reviewed recent concepts of iron compartmentalization between host and invading microbes. M. Luckey *et al.* (1972) showed that complex intermicrobic activities include the synthesis of different catechols

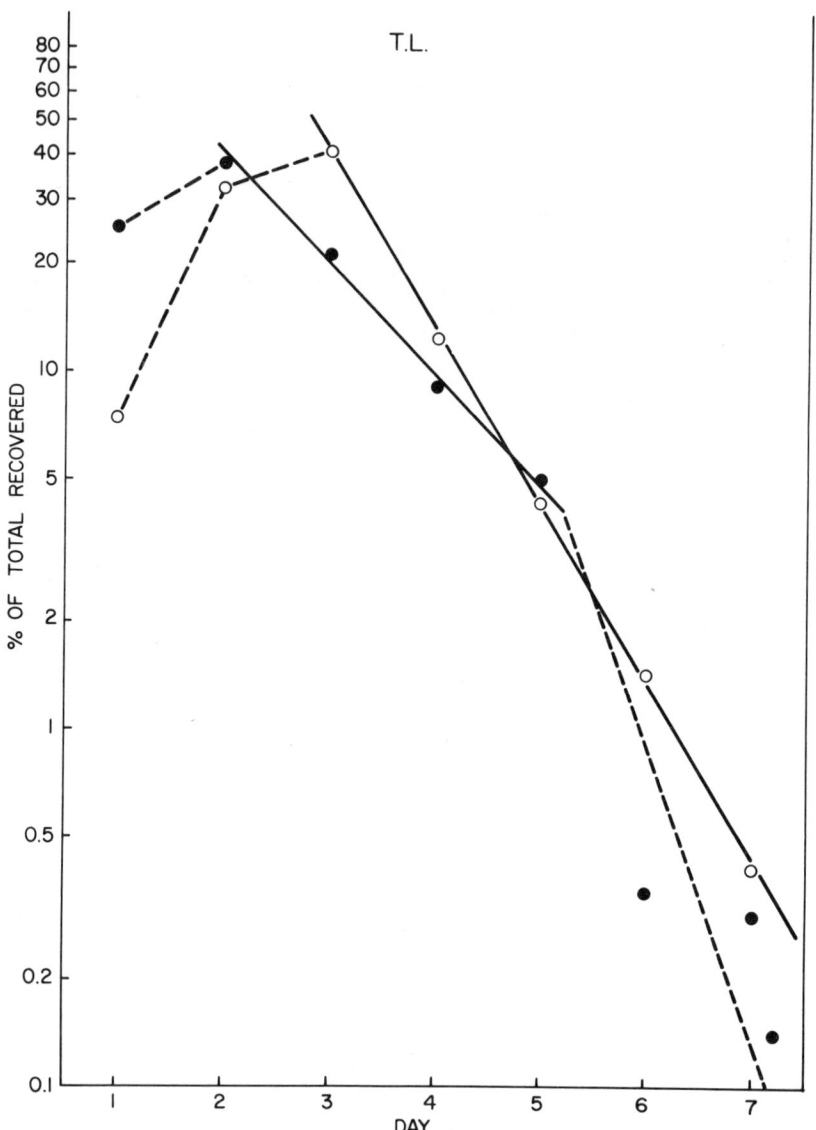

FIG. 2-7. Food passage through the alimentary tract. The two curves give preliminary data from a single marked bite of lunch with two different diets in a single subject. Neutron activation analyses of lanthanum were performed on feces by Don Dray of the Research Reactor, University of Missouri. Data from Luckey et al., 1977.

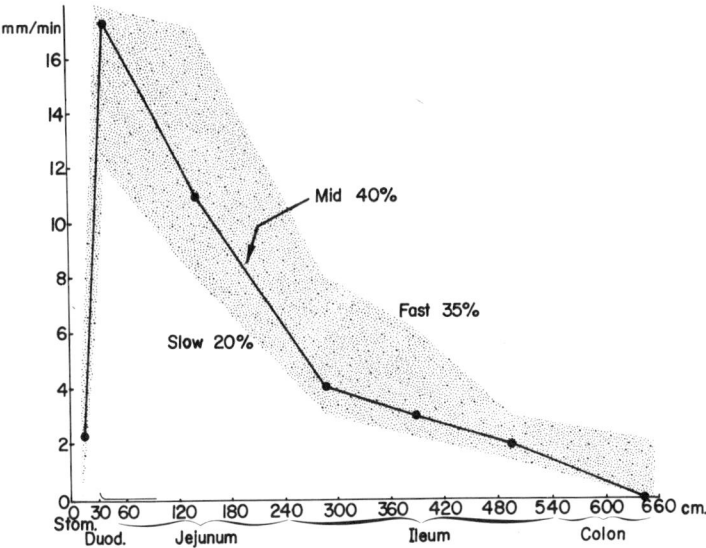

FIG. 2-8. Model flow rates of digesta in different parts of the gastrointestinal tract (from Luckey, 1972).

and hydroxamates as chelates and the competition for such iron complexes by different intestinal bacteria. A comprehensive treatise on microbial iron metabolism presents full details (Neilands, 1974). The iron-binding equilibria of different microbic siderochromes provide a competitive basis for one species to dominate others. This competition is represented in Fig. 2-9. It is difficult to imagine that the associated host mucosal cell is not equally involved in this competition for available iron. Any microbe or host cell that cannot utilize the iron siderochrome ligand produced by an associated microbe will suffer iron deficiency; those microbes that have permeases with an iron-binding equilibrium constant greater than the siderochromes of the media have a nutritional advantage for reproductive supremacy. The microbic production of siderochromes is a competitive action that will produce iron deficiency in those cells that cannot utilize such tightly bound iron. Microbes utilizing this survival mechanism have a great advantage. Mammalian mucosal cells do not appear to compete well in this activity; this may be the basic cause for anemia in our population, which seemingly receives adequate iron. If so, additional iron supplementation would be less helpful than control of the intestinal flora to reduce production of these ligands. Evidence for such activity was noted in mice fed Apollo diets (Luckey, 1970; Luckey et al., 1974). The survival of germfree mice was 80% at 2 months (Fig. 2-10A); survival of monognotophoric mice with

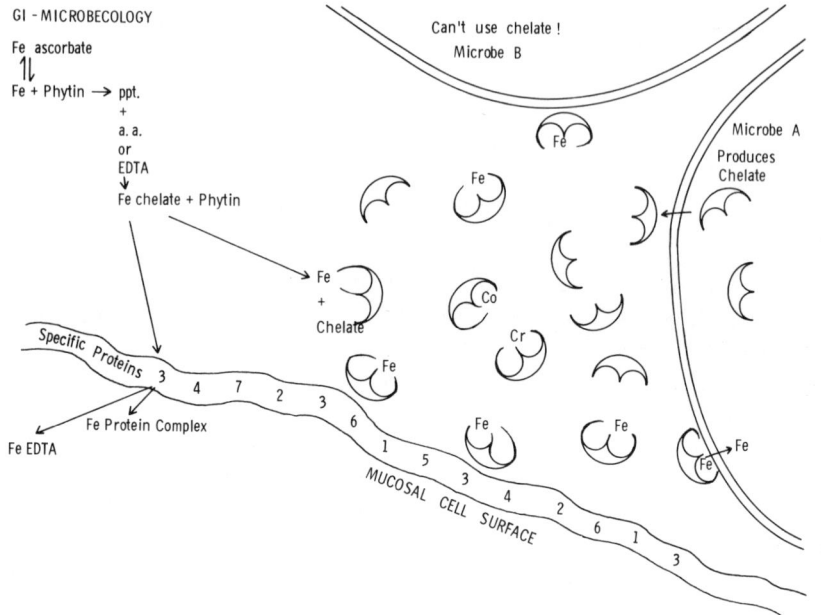

FIG. 2-9. Chelate competition within the intestinal lumen. This speculative concept illustrates the production of chelating compounds by one species of bacteria and the competition of other microbes and the host for the bound metals.

either *Escherichia coli* or *Lactobacillus leichmannii* was about 50%, but only 20% of mice di-associated with *E. coli* and *L. leichmannii* survived. These values showed a correlation with the average hemoglobin for the same groups (Fig. 2-10B). Since the hemoglobin varies greatly in anemic animals, it is assumed that anemia was a major contributing cause of death in mice represented by surviving littermates having an average of 6% hemoglobin. Similar reactions undoubtedly occur with other metals.

Absorption

The absorption or transfer of metal ions across the mucosa of the gastrointestinal tract depends on a number of factors in addition to the mode of uptake: (1) the physical–chemical character of the ion, (2) the physiologic pH in that segment of the gastrointestinal tract, (3) the time needed for the passage of digesta through the alimentary tract, (4) the microbic metabolism of the metal compounds, (5) the amount and composition of food eaten, (6) the presence of organic chelate compounds in the digesta, (7) interaction between the metal under consideration and metabolically similar metals and their compounds in the food, (8) biochemical mechanisms involved in the absorption of the metal ion (passive diffusion, active transport, or facilitated diffusion), and (9) the physiologic condition of the animal.

Negligible metal absorption occurs in the stomach. The major absorption takes place in the jejunum and ileum (Verzar, 1967). Rapid adjustment of osmotic equilibrium often makes net absorption minimal in the duodenum.

The major absorption of digested material occurs in the small intestine, with the majority of metals being absorbed in the mid-small intestine (Verzar, 1967). Although absorption is slow in the terminal ileum and colon, there is good evidence that absorption of many compounds occurs, and that there is ample time for absorption. These estimates are based on data that indicate that much of any one bit of food may be excreted 2–4 days later. Such data may be combined with average intake and secretion rates to provide estimates of the rates of flow through man's digestive tract

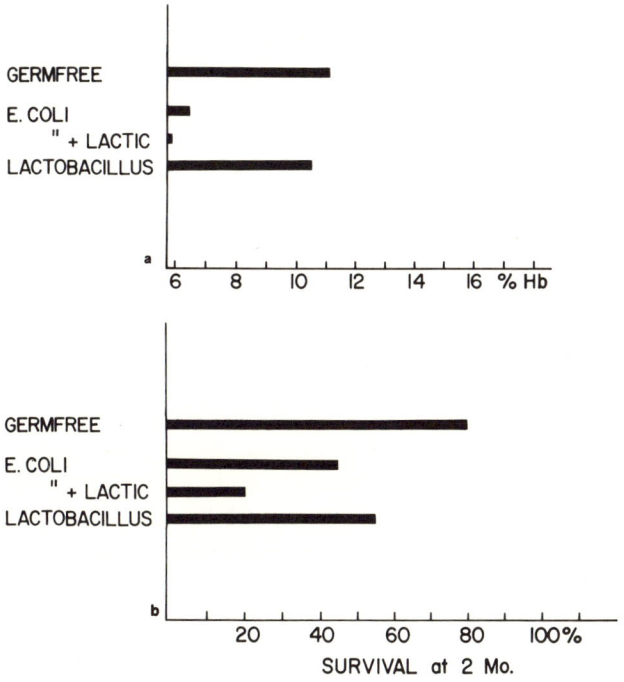

FIG. 2-10. Effect of two bacterial species on (a) hemoglobin and (b) survival of gnotobiotic mice.

(Luckey, 1974). In a human model, digesta remains 1–5 hr in the stomach, less than a minute in the duodenum, 1–6 hr in the jejunum, up to 1 day in the ileum, and an equivalent time in the colon. The terminal ileum may become a cyclic reservoir when the ileocecal valve is closed; digesta is held until released following evacuation of the colon or other stimulus. This reservoir may be relatively more important in man than in animals with functioning ceca. Unless the toxicant is cathartic, there is ample time for absorption; during this time, the chyme is intimately associated with villi as segmenting contractions force it back and forth in 5-sec cycles and villi contractions churn it as if in a micromixer.

The biochemical mechanisms involved in the gastrointestinal absorption of essential metals in mammals are mentioned in pertinent chapters whenever such information is available. Simple diffusion, active transport, and facilitated diffusion are the major mechanisms presented in standard texts and by Skoryna and Waldron-Edward (1971). The determining factors in absorption are: (1) lipid and water solubility, (2) size of the particle or metal ion, and (3) charge on the particle in relation to the charged pore size and potential across the membrane. In addition, specific proteins may be needed as carriers or regulators of absorption. The absorption of iron is influenced by a number of factors, one being the availability of ferritin in the mucosal cells (Jacobs and Worwood, 1974). Metals such as cobalt, nickel, and manganese are metabolically similar to iron, and compete for the same carriers at limited sites. Freeman *et al.* (1971) review the role of the microvilli in iron absorption. The calcium-binding protein of mucosal cells also carries strontium. Saturation of carrier proteins of the mucosa allows decreased absorption to be a major homeostatic mechanism for the reduction of metal toxicity by oral administration.

The physiologic mechanisms of absorption involve the columnar epithelium of the intestinal villi and have been reviewed by Sernka (1974) and Csaky and Autenrieth (1975). To be absorbed, material must pass through the glycoprotein–water matrix being continuously extruded by the cell. Much absorption presumably takes place at the base of the microvilli where much micropinocytosis is seen (Fig. 2-11). The outward-moving perimatrix sieve effectively prevents particulate matter from reaching the micropinocytotic sites at the base of the microvilli. Since absorption continues in the absence of pinocytosis, absorption must also occur in the microvilli. Diffusion, facilitated diffusion, and active transport must be greatly enhanced by the increased area provided by the microvilli and the exchange of fluid at the villus. Parsons (1975) reviewed the energetics of absorption into and exit from the intestinal mucosa. The brush border contains specific transport or carrier proteins, alkaline phosphatase, leucine

MODES OF INTAKE AND ABSORPTION 55

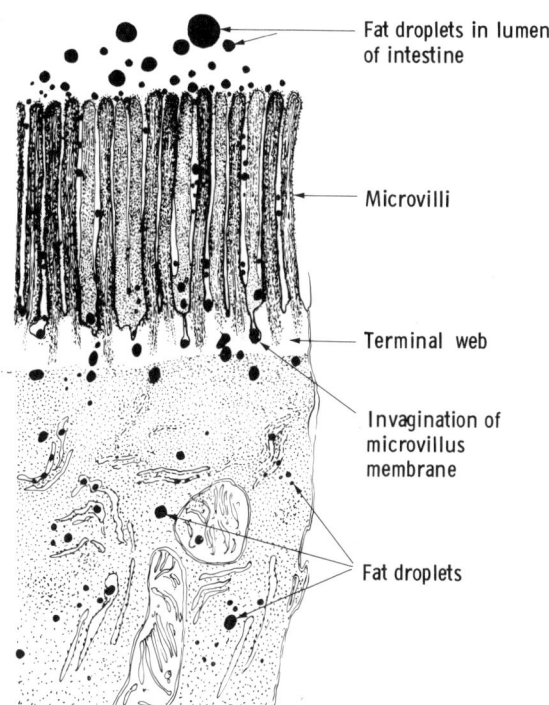

FIG. 2-11. The flow of small molecules is through the gel clathrate between the microvilli into the cell at the sides and base of the microvilli. Perimatrix secretion along the microvilli would provide a slow movement of large molecules and particulates away from the cell. Redrawn from Greep and Weiss (1973).

amino peptidase, and most of the disaccharidases. The glycocalyx may absorb other enzymes from the digesta flowing intimately past. Although man may secrete 3 liters of succus entericus, the villi of the small intestine absorb a total of about 8 liters per day. The villi contributions to the succus entericus (see Fig. 2-3) include: (1) mucoprotein from the perimatrix, (2) mucopolysaccharide from goblet cells, (3) blood neutrophils, (4) old mucosal epithelial cells, (5) their fragments, (6) enzymes and other material from the crypt epithelium and intercellular fluid that slips through tight junctions and cell extrusion lesions, and (7) wandering macrophages. Chyme flowing from the lumen through the perimatrix presents a continuously renewed supply of water-soluble materials to the absorptive and micropinocytotic sites of the cell. The space between the microvilli is 0.01–

0.05 μm. Appreciable quantities of particulate matter are not readily absorbed by this mechanism in healthy mucosa.

Transport of absorbed material through the columnar epithelial cell to the blood and lymph at the basement membrane has been shown with lipid stains. This reveals micropinocytosis as a major absorption mechanism for other material. In 1923, Kumaga (Verzar, 1967) noted that India ink and carmine were taken into epithelial cells by microendocytosis, with none being found elsewhere. This would suggest more absorption via endocytosis by mucosal cells than by macrophages.

Metals may become chelated by the mucoproteins that bind the apical desmosome. Cskay and Autenrieth (1975) suggest that particulates may pass through this tight junction between cells. Transport of ions across this junction and the wounds at the extrusion tips of villi account for the low electric resistance exhibited by the intestinal mucosa. The latter must account for most noncellular passage of ions as well as passage of large particles.

The percentage of absorption varies for each metal, and homeostatic absorption mechanisms may change these values according to the state of tissue saturation. The normal gastrointestinal absorption of individual metal compounds in physiologic doses is summarized here. Less than 5% of the following metals is absorbed: cationic aluminum, beryllium, bismuth, cadmium, lanthanides, gallium, indium, manganese, scandium, titanium, yttrium, zirconium, anionic gold, niobium, palladium, platinum, and tantalum. Absorption is about 15% for iron, 20% for calcium, 50% for magnesium and copper, and over 70% for anionic chromium, germanium, molybdenum, vanadium, selenium, cationic alkali metals, and thallium. Little absorption of metal compounds takes place during the few hours they are in the stomach. Histamine and test meals bring the acidity of the stomach of man and monogastric animals to pH 1-3, but a full meal neutralizes the gastric HCl to pH 5.5-6.5 (Rosenbaum et al., 1970). When most metal salts are introduced into an empty stomach, they are partially solubilized for limited absorption at the neutral pH of the intestines. Acid foods increase the absorption of metal salts.

The pH of digesta in the intestine usually ranges from 6.5 to 7.4; under these conditions the salts of Group III metals, soluble at acid pH, separate as insoluble hydroxides, and soluble salts of Group V metals become insoluble oxysalts. Heavy-metal ions react with sulfate, phosphate, oxalate, or phytate to form precipitates. Low-molecular-weight organic acids, amino acids, and peptides solubilize many metals. Free long-chain fatty acids react with metal salts to form insoluble soaps. Most of these insoluble complexes are eliminated.

Phagocytosis

Limited absorption of particulate matter occurs by macrophage phagocytosis. Macrophage phagocytosis is the major mechanism by which particulate matter is absorbed from a healthy gastrointestinal tract. As indicated in Fig. 2-1, macrophages exchange particulate matter of the alimentary tract, blood, and other tissues. This mechanism has been well documented in the lung, from which inhaled particles are transported into other tissues, including the alimentary tract. Tonsils are adapted for macrophage absorption (Payne, 1964). The phagocytosis of particulate matter 1–5 μm in diameter by tonsils is followed by drainage into the neighboring suprapharyngeal lymph glands. The total amount of absorption by tonsils may be negligible under normal conditions; however, when a toxic compound is absorbed, phagocytosis by tonsils is important. Azo dye particles are phagocytized in the duodenum (Barnett, 1959); polystyrene particles are phagocytized by intestinal cells for transport to lymphatics (Sanders and Ashworth, 1961). The early occurrence of nonmotile bacteria in Peyers patches at the initiation of the infection is added evidence for macropage activity. Freeman *et al.* (1971) indicated that macrophages may be significant in the transport of metal aggregates (i.e., ferritin) from the absorbing cell to other tissues. The continuous extrusion of perimatrix in the villi limits particulate absorption from the intestines.

Persorption

Persorption is the direct uptake of food particles from the digestive tract, and is another mechanism for particulate absorption. Caffein and prostigmine stimulate the activity of the musculature of the gastrointestinal tract and increase the persorption rate. These effects were studied in the mid-nineteenth century as the Herbst phenomenon; rediscovered by Hirsch (1906); confirmed in man, rabbit, dog, and cat by Verzar (1967); and extended experimentally by Volkheimer (1974*a*, *b*). Ingested starch grains are found in small quantities in blood, lymph, urine, muscle, brain, and all organs studied. Fisher (1931) noted that orally ingested yeast cells appeared in the portal blood of dogs; these studies have been verified by Krause *et al.* (1969). Although the quantities of particulate or nonparticulate matter found are extremely small, the overall effect cannot be ignored.

Persorption occurs at the extrusion zone on the tip of the villus (see Fig. 2-3). During the 2-day sojourn from the base to the tip of the villus, each cell may maintain regular relationships with each neighbor. As the cells round the tip of a fingerlike villus, previous associations (the tight junctions at the terminal web, desmosomes, and interdigitations) become disrupted, and the cell becomes a less integral component of the villus. Thus, the wound at the extrusion zone is more complex than the simple loss of a few cells in the overcrowded tip. As the villus contracts, internal pressure is added to the constant cell migration pressure, and old cells erupt from the tip. The cell-migration pressure comes from continual replacement of about 90 cells around the circumference of a villus by cell reproduction at the villus base. During the few seconds following contraction, the relaxation of the villus to its full length reduces the internal pressure and provides conditions for a microimplosion of particulates through the villus tip wound at the extrusion zone. Particulates and associated material are introduced directly into the lacteal system via persorption.

The quantity of particulate matter that enters the body by either phagocytosis or persorption is very small in mature, healthy mammals. The effect would be significant only with highly toxic or cumulatively toxic metals.

Gavage

Gavage is a modified form of oral intake with the transfer of material into the stomach through a tube. Some compounds, such as the dyes Fd and C Red No. 2, are harmless when given to test animals in their diet, but are toxic by gavage. Warden and Harper (1964) and Boyd (1972*a*) confirm the concept of greater toxicity of drugs given by gavage as compared with dietary administration; the acid of the empty stomach may hydrolyze or activate the drug into a more potent form.

Aboral Administration

Rectal administration is utilized for drugs that may be seriously changed by digestive enzymes. Rapid absorption may be obtained for many compounds that would be inactivated by the liver and hence would not be efficiently utilized *per os,* since absorption via the portal vein would take them directly to the liver. Suppositories or enemas are used to give drugs that would cause vomiting. Absorption is enhanced by either aqueous or lipid solubility of the drug or toxicant. Sollman (1948) suggests that the simultaneous administration of alcohol or ether will increase the absorption of rectally introduced material. Retention time may be very short unless physiologic manipulation or anal block is applied. Coldwell *et al.* (1969) noted a similar rate of absorption of soluble drugs given orally or by suppository. Boyd (1972a) concluded that readily absorbed drugs are equally toxic when given orally or held rectally for 2 hr.

INHALATION

The Respiratory Tract

The respiratory tract has three main regions: the nasopharyngeal, pulmonary, and alveolar (Fig. 2-12). The pulmonary region consists of the trachea, two main bronchi, numerous generations of cartilagenous bronchi, and multiple bronchioles ending in alveolar ducts. Ciliated epithelial cells predominate over goblet cells in the nasopharyngeal cavity and the trachea, becoming progressively less numerous in the terminal bronchioles. Alveoli are devoid of ciliated cells. The epithelial lining of the nasopharyngeal cavity, trachea, and bronchi secretes mucus; most irritants excite increased mucus secretion. Alveolar epithelial cells efficiently absorb soluble metal compounds; goblet and other cells lining the trachea and bronchi absorb these compounds poorly.

During inhalation the air velocity is high in the nasopharyngeal region. The directional change of inhaled air is very abrupt in the nasopharyngeal region, and progressively less abrupt in the trachea and bronchi. Air moves by diffusion in the bronchiolar and alveolar regions. Air velocity and its directional change influence the aerodynamics of the respiratory tract (Davies, 1961; Hatch and Gross, 1964). The decreasing velocity of blood and air movement as the vessels become smaller and the surface area increases is augmented by muscular activity in bronchial and alveolar ducts and elastic tissue throughout the entire respiratory tract. Representative tidal and alveolar volumes for different species are given in Table 2-1.

The gross anatomic structure of the respiratory tract prevents large particles from reaching the alveolar region of the lung. Gaseous and microparticulate (< 0.2-μm diameter) absorption occurs primarily in the 3×10^8 alveoli of man; this ever-moving surface of 30 m^2 (100 m^2 in deep inspiration) equilibrates with 10 m^3 air per day. The alveolar capillary bed contains a network of blood vessels that total 2000 km. Hatch and Gross (1964) have characterized particulate deposition in lungs.

Inhaled air is humidified in the nose, pharynx, and mouth to 100% relative humidity. The mucus of the respiratory tract forms a blanket between epithelial cells lining the tract and the air and particles respired. It is produced by mucus glands, goblet cells, and Clara cells (Kilburn, 1967).

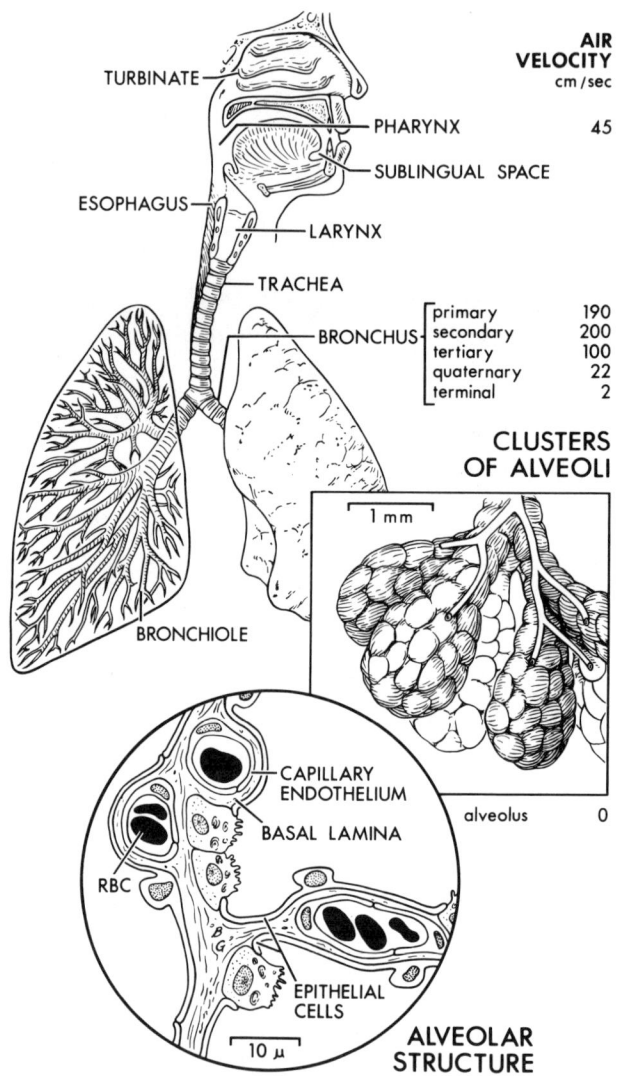

FIG. 2-12. Diagram and air flow rates of the respiratory tract.

Highly soluble material may complex with the water vapor or may be readily dissolved in the respiratory mucus. This thick extracellular fluid contains phospholipids that reduce surface tension and change the absorptive properties of the fluid. A variety of irritants increase the secretion, flow, and discharge of this moving blanket. Allergies and certain disease

TABLE 2.1 Minute-Volume of Air Intake[a]

Species	Condition	Body weight (kg)	Tidal volume[b] (liters/min)	Alveolar volume[b] (liters/min)
Man	Rest	70	7.4	5.2
	Light work	70	28.6	20.0
	Rest	54	4.5	3.2
	Light work	54	16.3	11.4
Horse	Standing	500	200	140
Cow	Standing	400	100	70
Pig	Standing	200	37	26
Sheep	Standing	60	7.1	5.0
Dog	Standing	20	5.2	3.6
Cat	Rest	4	1	0.7
Monkey, rhesis	Rest	3	0.86	0.60
Guinea pig	Rest	0.5	0.16	0.11
Rabbit, domestic	Rest	4	0.82	0.57
Rat, white	Rest	0.3	0.073	0.051
Hampster	Rest	0.1	0.06	0.04
Mouse, white	Rest	0.03	0.024	0.017

[a]From Altman and Dittmer (1974).
[b]Assume that tidal volume × 0.7 = alveolar volume.

states cause conditions considerably different from normal. Deep or shallow breathing produces different particle deposition patterns.

Boyd (1972b) characterized the purposes of respiratory tract fluid as (1) an essential protection against dehydration of sensitive mucosal cells, (2) a homeostatic component necessary for cilia action, (3) an essential for prevention of particulate penetration to cell membranes, (4) alleviation of viral and bacterial infections by both mechanical removal and antibody participation, (5) removal of inert foreign particulates, and (6) a vehicle for removal of internal toxicants. This fluid is analogous to the epidermis, continually shedding entrapped material from the tissue cells adjacent to it; its contents sometimes include erythrocytes and regularly include sloughed epithelial cells and bacteria. Estimates of the volume of respiratory tract fluid in a variety of laboratory animals (Table 2-2) suggest that the flow in man would approximate 2 ml/kg per day, or about 200 ml/day for an average adult; Toremalm (1960) indicated that the quantity may be 0.5 ml/kg per day. The flow is inversely related to relative humidity. The quantity secreted increases dramatically when the respiratory tract is disturbed by sensory, psychic, particulate, or chemical stimuli. Thus, the respiratory tract fluid is a moving barrier about 5 μm thick that protects our living

TABLE 2-2. Volume of Respiratory
Tract Fluid[a]

Laboratory species	Tracheal fluid[b]
Mouse	
Rat	7.7
Hampster	
Guinea Pig	2.5
Rabbit	2.08 (3.2)
Cat	2.28
Dog	0.58
Pig	
Monkey	5 (est.)
Chicken	2.6–8.4

[a]From Boyd (1972b), with high and low values deleted from the calculation.
[b]ml/kg body per day.

tissues from an increasing variety of noxious airborne agents. Small particulates trapped in mucus of nasal, tracheal, or bronchial epithelial cells are moved by cilia to the glottis, and are usually swallowed to become digesta of the alimentary tract. Repiratory tract fluid is also recovered by absorption into the cells lining the tract, by phagocytosis, and by seeping into the lymph (Boyd, 1972b).

The respiratory fluid is moved by about 200 cilia per epithelial cell. These move small particles up to 1 cm/min by beating 20 times/sec; the rate is only 1 mm/min in bronchioles. Single particles may travel from bronchioles to larynx in 30 min. The cilia average 0.3 μm in diameter and 6 μm in length (Fig. 2-13). Few drugs affect their activity, but they are sensitive to temperature and humidity changes (Boyd, 1972b). Coughing accelerates movement to the pharnyx.

Blood–gas exchange (as used here, "gas" includes particulate matter) occurs by a series of events beginning with a respiratory tract fluid–gas exchange in which diffusion, hydration, and solubility must play important roles. Material incorporated into this special fluid establishes a second equilibrium between cell and interstitial fluid prior to the penetration of material into the capillary blood or removal via the lymphatics. The time for this total exchange is negligible for practical considerations; the physical–chemical characteristics of the membranes act to speed the absorption of soluble compounds much faster than that of less soluble gases or particulates. Following absorption in the lung, there is a fast distribution to blood-rich organs (1–2 min), slow distribution to other organs and fat depots (10–20 min), and a very slow equilibration to vessel-poor tissues such as cartilage.

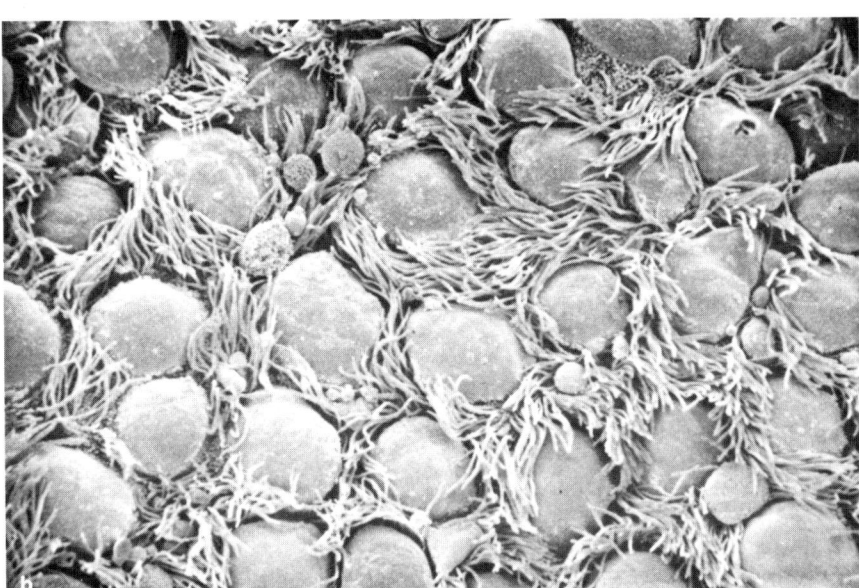

FIG. 2-13. Scanning electron micrographs of respiratory cilia: (a) Horse bronchiolar epithelium. Note the short, blunt microvilli covering the Clara cells and the longer microvilli at the base of the ciliated cells (field width ~20 μm). (b) Horse bronchus ciliated and nonciliated (mucous) epithelial cells. The apical surfaces of the mucous cells are fringed with microvilli. Openings at the apex of some mucous cells indicate the discharge of mucigen (field width ~70 μm). (Contributed by Dr. Janice Nowell, E. M. Facility, Thimann Laboratories, University of California, Santa Cruz.)

The respiratory tract is also an effective excretory organ for volatile compounds such as ammonia, low-molecular-weight ketones and ethers, and selenium oxide. Although small concentrations may be involved, the continuous movement of air through the lungs enhances contact with a vast capillary bed and ready diffusion into air as a homeostatic mechanism for volatile metal salts.

The distribution, deposition, and retention of inhaled material in the pulmonary tissues and its eventual transfer to other tissues depends on the physicochemical properties of the material, the aerodynamics in the respiratory tract and lungs, and the initial responses of the pulmonary tissues to the toxicant. Subtle effects from toxicants reaching the lung or primary target tissues cannot usually be evaluated accurately, because previous trauma to the respiratory tract under intermittent exposure to mixtures of chemicals and microbes allows no judgment of what caused the changes. Tanami (1967) used germfree animals in a clean environment to avoid these problems; the lungs of germfree animals readily reveal subtle, primary changes in the presence of specific airborne toxicants. In healthy individuals the tracheobronchial tree is usually free of microorganisms (Lourenco, 1970). The resident microflora of the upper respiratory tract have no known direct function in the inhalation toxicity of metals. Evidence from studies of germfree animals suggests that numerous recurrent microinvasions of the microflora stimulate the development of cilia and the production of leukocytes and leave microscars; therefore, the microflora probably play an indirect role (Luckey, 1963).

Inhaled Pollutants

Airborne vapors and particulates, called *aerosols,* include organic and inorganic compounds of natural and synthetic origin. Air of the urban atmosphere contains potentially toxic salts of metals such as cadmium, lead, antimony, selenium, thallium, vanadium, nickel, and zinc. Compounds of zinc, lead, cadmium, arsenic, antimony, chromium, and mercury are "enriched" in polluted urban air up to 1000-fold above the average levels at which these metals occur in the earth's crust (Natusch *et al.,* 1975). Most toxic components of air pollution emanate from automobile exhausts, coal-fired power plants, garbage incinerators, metallurgical and refinery operations, and aerosol cans of cosmetics, pesticides, paints, varnishes, and propellents. Lung damage from cigarettes, however, is far greater than that from these pollutants (Bates, 1972). Unfortunately, this applies as well to the nonsmoker who is in proximity to a smoker. Bates summarized the possible damage to the respiratory tract (Table 2-3) and

TABLE 2-3. Possible Effects of Respiratory Tract Irritants[a]

	Major bronchi	Terminal bronchioles	Alveoli
Short-term:	Inhibit deep breathing	Increased susceptibility to injury	Increased susceptibility to injury
	Reverse bronchiospasma		
Long-term:	Cilia paralysis	Diminished defenses	Increased cells and macrophages
	Hypersecretion of mucus	Adverse effect on surfactant	Release of proteolytic enzymes
	Mucous gland hypertrophy and extension to smaller bronchi	Goblet cell metaplasia	Possible alveolar destruction and emphysema
	Increased infection susceptibility	Premature closure of airways	Increased infection susceptibility
	Chronic cough	Decreased gas exchange	
		Release of proteolytic enzymes	

[a]From Bates (1972).

emphasized the potentiation of one harmful material by another. Cigarette smoke contributes nickel, chromium, cadmium, lead, and aluminum compounds to the inhaled air. Although some of these pollutants are gases, the majority are particulate. Man and animals are exposed to these pollutants more through inhalation than through other modes of intake. Most metal compounds, while exerting some transient or secondary toxic effects in the lung, produce toxic effects at primary sites elsewhere in the body. Some metals (nickel and chromium) and their compounds produce pulmonary neoplasms.

The fumes from burning coal and petroleum products contain trace amounts of most metals. Lead, thallium, antimony, selenium, arsenic, nickel, chromium, and zinc are adsorbed onto the surfaces of fine particles within the fumes (Natusch el al., 1974, 1975); the smaller the particle size, the more metal adsorbed and the greater the concentration of the adsorbed metals. This dependency on particle size is not evident for the following minerals: bismuth, tin, cobalt, iron, manganese, vanadium, titanium, calcium, potassium, magnesium, aluminum, and beryllium. Metals are adsorbed on the surfaces of the particles, where they can readily react with tissues on which the particles impinge. Environmentalists have noted that automobile emissions are fine particulates, but that fumes from commerical plants carry relatively large particulate matter. The particle sizes of airbone pollutants are given in Fig. 2-14.

In a chemically polluted atmosphere, metal compounds occur as either vapors or particulate matter. Certain metals, such as mercury, and metal oxides, such as arsenic and selenium oxides, can exist as vapors; 80 $\mu g/m^3$ selenium as selenium oxide and 70 $\mu g/m^3$ arsenic as trioxide can exist as vapors at 25°C. The selenium oxide is unstable and reacts with SO_2 or other compounds (Frost, 1972). Metal compounds are usually concentrated or adsorbed on the surfaces of colloidal particles; in some cases, they are distributed throughout the aerosol. The concentration of such toxicants increases with decreasing size of the metal particle and of the aerosol colloid to which it is attached. High-temperature combustion sources emit a proportionally large number of extremely small toxic particles, especially the metal oxides; the higher the temperature, the smaller the particles and the greater their adsorptive surfaces (Hatch and Heneor, 1948). Metals, especially volatile compounds, are more toxic when concentrated on the surfaces of the microscopic particles (about 0.1 μm in diameter) than they would be when dispersed as larger particulates. Casarett and Doull (1975) note that particles 0.2–0.5 μm in diameter are the most stable in aerosols. The effects from electrostatic charge increase as the particle size decreases. Reactivity and surface area increase tremendously as particle size

MODES OF INTAKE AND ABSORPTION

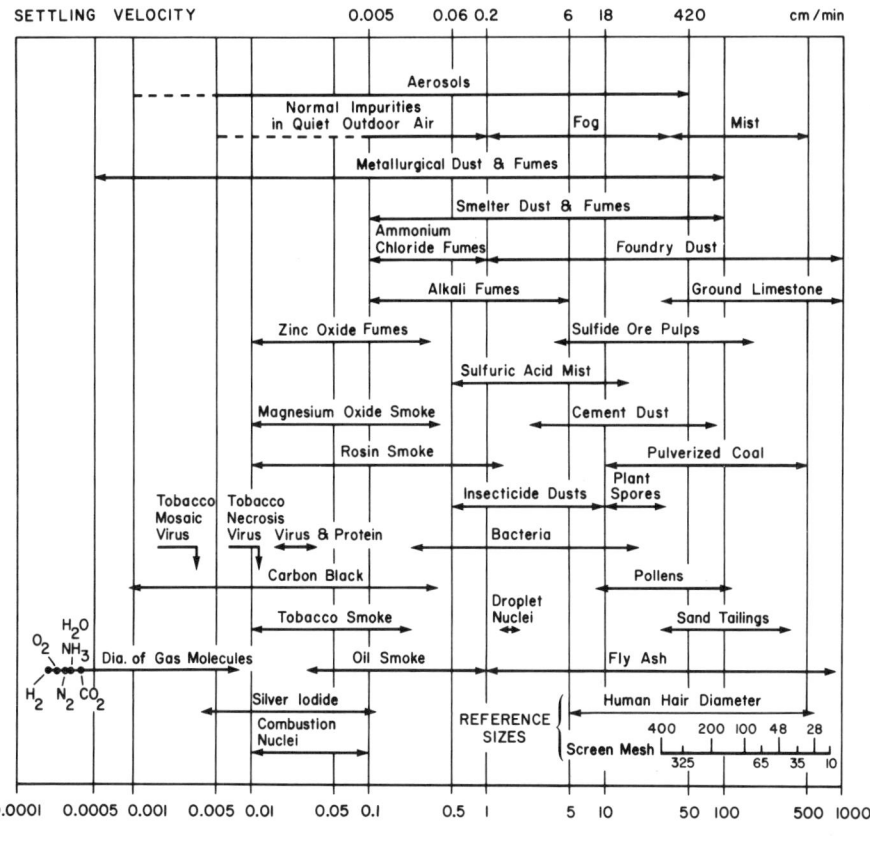

FIG. 2-14. Particle sizes of airborne pollutants.

decreases. Organically bound metals seem to have a special affinity for cell membranes.

The physical diameter or shape is not a suitable measure of the behavior of particles in the respiratory passage. The aerodynamic diameter, a function of both physical diameter and density, serves to characterize the particle. Between two particles of the same dimensions, that with higher density is considered to be a larger particle on an aerodynamic basis. The size of a hygroscopic particle changes markedly on entry into the water vapor-saturated respiratory tract. In these discussions on inhalation toxicity, the particle size invariably refers to the aerodynamic diameter—which is usually expressed as count median diameter (CMD) or mass median diameter (MMD)—along with a distribution function and the geometric standard deviation. Filamentous particulates have been less well characterized. Their deposition is related to diameter, length being relatively less important.

In the discussion of particulate deposition, uptake, and removal, it is assumed that the respiratory system is normal. Pollution, drugs, and lung diseases may dramatically change the morphology, chemistry, and physiologic reactions. Also metal toxicity can change lung function (Bonnell, 1955).

Particulate Deposition

The deposition and distribution pattern of inhaled toxicants depend on respiration characteristics and particle shape, size, and solubility. An excellent review of particulates in the respiratory system is presented by Casarett and Doull (1975). The amount absorbed or exhaled depends on the total dosage and the physicochemical character of the particle, as well as on the condition of the individual.

Material inhaled through the nose enters a wind tunnel labyrinth in which the high velocity of the inspiratory air, with its rapid and abrupt change of direction, forces particulate material against mucus-covered turbinates. There is increased deposition of particles at the points of bifurcation of each tube (Natusch et al., 1975). The settling velocity due to centrifugal force is high for large particles (5–30 μm diameter), and most are trapped in the nasopharyngeal region by inertial impact. Smaller particles sediment from gravity. Diffusion characterizes the activity of very fine particles which do not sediment at rates significant to respiration; they exhibit Brownian movement. If the impacted particles or droplets of vapor are water-soluble, they are incorporated into the mucus and may be absorbed into blood. Irritation initiates greater mucus secretion. The trapped material with mucus may be expelled through the nares into the pharynx for swallowing or external discharge.

Particles 20 μm in diameter settle at a rate of about 1 cm/sec. This rate, plus their large size, would rarely allow them to penetrate the nasal labyrinth at normal respiration rates, 6–12 breaths per minute. Particles 2–10 μm in diameter settle at about 0.05–0.3 cm/sec and are virtually all deposited in the nose, pharynx, and trachea. Particles less than 0.1 μm in diameter show virtually no deposition in the respiratory tract and resemble gases in this respect.

Particles 0.1–2 μm in diameter are of special interest in pollution. Some particles of this size penetrate the trachea and bronchi of the pulmonary system and are gradually trapped onto the mucous lining. Small particles progress slowly in bronchi due to the increased air velocity and less abrupt directional changes of air flow. As described above, the epithelial lining of the trachea and bronchi also secretes mucus and has cilia that trap and transport the insoluble particulates to the pharynx, where most is swallowed, and some is expectorated. Soluble material is absorbed. The bronchiolar regions trap particles ranging from 1 to 5 μm in diameter, and some may be expelled in the exhaled air. Microscopic particles less than 1 μm in diameter may also settle in the tracheobronchial region and are transferred to the blood. These fine particles are subject to electrostatic forces. Particulate matter that reaches the alveoli is usually less than 1 μm in diameter. In the alveoli, the air velocity is minimal and the directional change of air flow is negligible. Particles deposited in the alveoli are mostly cleared from the pulmonary parenchyma by alveolar macrophages through phagocytosis. Some metals are toxic to these cells (Waters *et al.*, 1975). Very fine particles less than 0.1 μm in diameter can diffuse through the alveolar epithelium into lymphatic vessels and finally into the blood. Their setting time, however, is long. Inhaled metallic vapors and particulates of submicroscopic size may be absorbed into the blood to the extent of their solubility; the major portion of insoluble microscopic particles will be retained or exhaled, as shown in Table 2-4. Goldstein *et al.* (1974) present a discussion of the inhalation of drugs. Some metals, such as aluminum, accumulate in the lungs with age. During a lifetime, the accumulation of residual particulate matter is considerable. Deposited insoluble particles may be sequestered in tubercles at the expense of the alveolar space, where they are subjected to very slow biochemical reactions and cell erosions and are transformed into soluble ionic forms.

Using insoluble nonhygroscopic particles as models, illustrative data are reproduced in Tables 2-4 and 2-5, with sizes characteristic of urban aerosols. In a large city, one may expect 1 mg particulates/m^3 air. This amount would account for about 10 mg/day, 2 g/year, or 200 g/lifetime. These estimates for pulmonary deposition of insoluble particulates are minimal and might be doubled according to Hatch and Gross (1964). The material not deposited is assumed to be exhaled. A composite of their data with that of Palm *et al.* (1956) is given in Fig. 2-15. It is emphasized that larger particles are deposited in the nasopharyngeal region, where cilia action and swallowing would take them into the stomach, or expectoration or nasal discharge would eliminate them. Soluble material should be retained much more than these data suggest.

TABLE 2-4. Percentage Retention of Inhaled Particles in Different Regions of the Respiratory Tract[a]

Region	Particle size:	20 μm	6 μm	2 μm	0.6 μm	0.2 μm
			Retention (%)			
Mouth		15	0	0	0	0
Pharynx		8	0	0	0	0
Trachea		10	1	0	0	0
Pulmonary bronchi		12	2	0	0	0
Secondary bronchi		19	4	1	0	0
Tertiary bronchi		17	9	2	0	0
Quaternary bronchi		6	7	2	1	1
Terminal bronchioles		6	19	6	4	6
Respiratory bronchioles		0	11	5	3	4
Alveolar ducts		0	25	25	8	11
Alveolar sacs		0	5	0	0	0
TOTALS		93	83	41	16	22

[a]Estimates of Hatch and Gross (1964) based on 450 cm^3 tidal air and 4-sec cycle with 300 cm^3 air/sec.

The deposition of insoluble particles in the respiratory tract projected by Hatch and Gross (1964) (Table 2-6) is considerably less than Palm et al. (1956) reported for clay dusts. The latter investigators found that guinea pigs retained more particles in the upper respiratory tract than had been projected for man. About 90% of all particles over 1 μm in diameter and 50-80% of particles 0.2-1.2 μm in diameter were retained. Since the settling time of particles less than 1 μm in diameter is long compared with respiration time, these small particles may be treated as gases.

The high retention of particles suggests a rationale for the conversion of inhalation data from an air concentration and time to a milligram basis. When inhaled in small quantities, powders greater than 10 μm in diameter are deposited in the upper respiratory tract and removed to the alimentary tract. Particles 2 μm or more in diameter are deposited in the bronchi and

TABLE 2-5. Deposition of Metals in the Respiratory System[a]

Metal	Mass median diameter (μm)	% Deposited by region		
		Nasopharyngeal	Tracheobronchial	Pulmonary
Iron	2.7	48	7	22
Lead	0.56	17	6	32

[a]From Natusch et al. (1974), who give percentage of material deposited compared with the total inhaled.

MODES OF INTAKE AND ABSORPTION

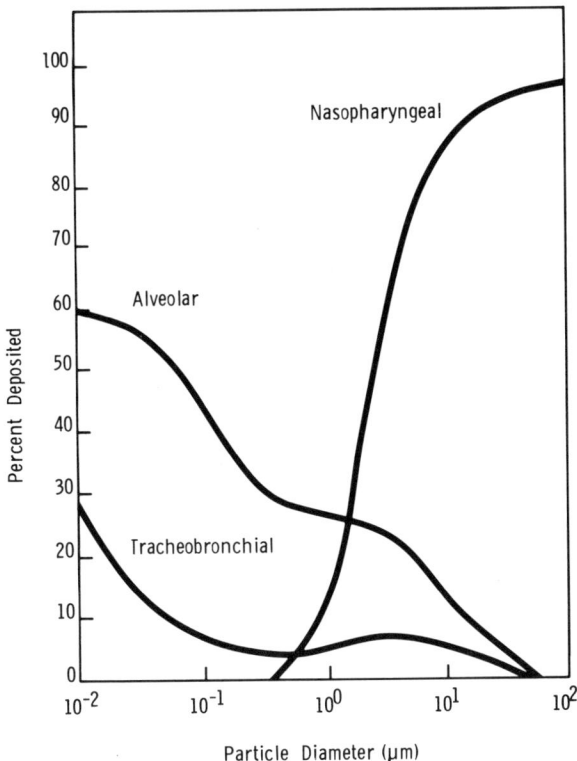

FIG. 2-15. Retention and deposition in man of inhaled particles of different sizes (Palm et al., 1956).

trachea. Little correction would be needed to convert the total quantity of particles greater than 2 μm in diameter inspired to total dosage. The difference between 90 and 100% deposition can be ignored on the basis that the larger particles have less physiologic action. Insoluble particles 0.2–2 μm in diameter form a set in which 63–80% are trapped in the pulmonary system, and about one-third would be exhaled. The alveolar volume (70% of inspired air) could therefore be taken as the dosage for particles less than 2 μm in diameter. Except for the 30% of each inspiration that remains in the upper respiratory tract, fine particles less than 0.2 μm in diameter are completely available for absorption in aveoli. Of the total fine particles inspired, therefore, 70% are available to the alveoli; this is the alveolar volume. This reasoning indicates that in dosage estimations, the alveolar volume should be used for particles less than 2 μm in diameter, and is

TABLE 2-6. Approximate Deposition of Inhaled Insoluble Particulates[a]

Material	Particle size (μm)	% Deposited by region[b]		Approximate total
		Upper respiratory	Deep pulmonary	
Gas	<0.1[c]	0	>60	60
Fumes	0.1[c]	0	~60	60
	0.2	0	53	53
	0.4	0	76	76
	0.6	0	80	80
Smoke	1	3	53	56
	2	60	26	86
Dust	3	70	20	90
	5	85	15	100
	10	90+	<10	100
	20[c]	100	0	100

[a] From Hatch and Gross (1964).
[b] The remainder is not found in lungs at the end of the experiments; it may be either exhaled particulates, absorbed microcolloids, or dissolved material (see Fig. 2-15).
[c] Estimated from Fig. 2-15.

appropriate for particles greater than 2 μm in diameter. These guidelines will be used to relate inhalation data to data from other modes of administration.

Particle Uptake and Removal

Microscopic particles may be trapped in the alveoli and phagocytized by macrophages, which then transport them through tissues to the lymphatic system or the digestive tract. Many microscopic particles from the lungs are added to the digesta of the alimentary tract; some go via the pharynx, some are transported by macrophages directly to the alimentary tract, and some enter the liver and then enter the enterohepatic excretion.

Natusch et al. (1974) estimated the absorption of lead from their data (Table 2-5). Assuming that 70% of the lead in the pulmonary region is absorbed and that only 10% of the lead from the upper respiratory tract is absorbed after it is swallowed, 22% of inhaled lead is absorbed from the lungs while 10% is absorbed from the alimentary tract. They suggest that the average urbanite obtains about 33% of his total lead intake each day (30 μg) by inhalation.

The lung is an effective excretory organ for volatile metals such as Se_2O_3. Although small concentrations may be involved, the continuous movement of air through the lung enhances contact with a vast capillary bed and makes exhalation an effective homeostatic mechanism for volatile metal salts.

Removal of insoluble particles involves physical action. Nasal epithelial cilia propel particulates and water-insoluble droplets into the pharynx, where they may be either expectorated or swallowed. Experiments indicate that over 90% of such particles move into the alimentary tract in 1 month; only 1–5% may remain in lung deposits.

Removal of fine particles occurs in distinct phases (Hatch and Gross, 1964; Lourenco et al., Gamsu et al., 1973; Casarett, 1972; Casarett and Doull, 1975). The rapid phase has a half-life of about 2 hr (Albert et al., 1967); it is completed within 6–24 hr. Particles are removed from the upper respiratory tract by coloidal movement of the mucous blanket at a rate of 1 cm/min. Tracheal material is completely removed within 20 hr by healthy respiratory cilia, according to tantalum insufflation studies by Gamsu et al. (1973). Much of the insoluble metal oxides that have been used to explore this phenomenon is excreted unabsorbed in the feces. Lourenco et al. (1971) found that smokers retain considerably more of the 2 μm particles than do nonsmokers during the first 3 hr; 24 hr following inhalation, the retention was comparable in the two groups.

Gamsu et al. (1973) also found that clearance time increased dramatically with decrease in size of bronchi. Proximal bronchi showed good clearance in 3 days, and terminal bronchi showed no clearance for 4 days; after that time, the rate of clearance equaled that of the proximal bronchi.

Material that is not rapidly removed from the lungs has very different clearance characteristics. Much is usually deposited in the terminal bronchioles and the alveoli. The intermediate phase (phase II of Casarett) is slow but definite. Removal may take 1–2 months. Presumably, this deposited material accumulates a variety of cells and debris. Most of this accumulation can be mobilized in healthy lung tissue that is not overwhelmed with too much material. Macrophage clearance at a slow, constant rate characterizes phase II. Continued accumulation, however, might start a granuloma or a fibroma. The half-life of this intermediate clearance of deposited particles may extend from 1 month to several months, depending on the quantity and characteristics of the particles.

The material deposited in deep pulmonary regions is mostly less than 0.1 μm in diameter and becomes more readily soluble or moved by virtue of its colloidal properties. The lymph drainage of this area of the lung is effective. Macrophages readily transport fine particulates during the first few days following exposure. It is assumed that few particulates penetrate bronchiolar or alveolar tissue prior to being phagocytized; few may travel in the lymphatics or through tissues as naked particles.

The third phase of removal involves material that has been effectively sequestered from normal metabolic activities. Gamsu et al. (1973) found no movement for 15 months from tantalum in alveoli. Insoluble metal compounds and colloids with heavy metals with macromolecules to form a stable, unreactive material, i.e., a denatured protein. This long-term clearance predominates when large quantities of particulates less than 1 μm in diameter are inhaled.

In addition to the removal systems described above, the very slow dissolution of "insoluble" materials, such as silicates, continues as long as there is effective interface between the particulate material and living cells. This action decreases if an effective tubercle is formed around the particulate. This slow phase IV activity may continue for years.

Drugs, infections or cell diseases, presence of irritants, and desquamation may dramatically change these clearance patterns. Delay in removal is diagnostic of an overwhelmed tissue or a pathologic condition.

Absorption

Inhalation provides an effective port of entry for toxic metals to reach organs and tissues without previous processing by digestion in the alimentary tract or by detoxication in the liver. Metals are absorbed 10 times more effectively by the lungs than by the intestines. In humans, the total surface area of the lungs is about 35 times greater than that of the skin.

About 70% of each breath reaches the alveoli, where histologic structures and the slow movement of respiratory gas and alveolar capillary bed provide effective contact for gaseous exchange. Soluble gases, such as NO_2 and O_2, and metals, such as thallium, are absorbed in the alveoli in healthy lungs; some, such as SO_2, are readily absorbed in the large airways. Small quantities of soluble vapors, such as acetone, are quickly and almost completely absorbed.

The relative amount of material absorbed in the respiratory tract, especially in the bronchioles, decreases with lung dysfunction. Prolonged inhalation of fine dust may considerably decrease the total absorptive capacity of the lungs. When industrial dusts accumulate in the bronchioles, the irritation reaction may include formation of fibrosis with subsequent dysfunction of that portion of the lung. This fibrosis portends increased toxicity. Partial blockage tends to decrease pulmonary circulation, making the bronchial circulation important in absorption. Material absorbed in the bronchial arteries and veins goes directly from the arterial blood to tissues without benefit of the slow filtration through the reticuloendothelial components of lymphatic or pulmonary capillaries. The pulmonary blood is

Physiologic Effect of Particle Size

In inhalation toxicity, the size of the particulate matter is a most important feature. Implications for differential toxicity of the same quantity of a metal adsorbed onto a fine or a coarse microscopic particulate are evident. The former is deposited farther down the respiratory tract than the latter, and is much more harmful than the latter even when the same quantity of metal is involved. Flinn et al. (1940) noted that inhalation of a coarse dust of manganese oxide was asymptomatic, while pneumonitis resulted when a fine dust was inhaled. Dygert et al. (1949) found that the toxic action of uranium oxide was not evident until particles less than 3 μm in diameter were used (Fig. 2-16). The larger particles were inactive; the small particles provided enough surface area to be highly active colloidal particles.

As shown in Table 2-4, if the particle size becomes too small, the aerosol is readily exhaled, and absorption depends on the water solubility of gases and fine particulates.

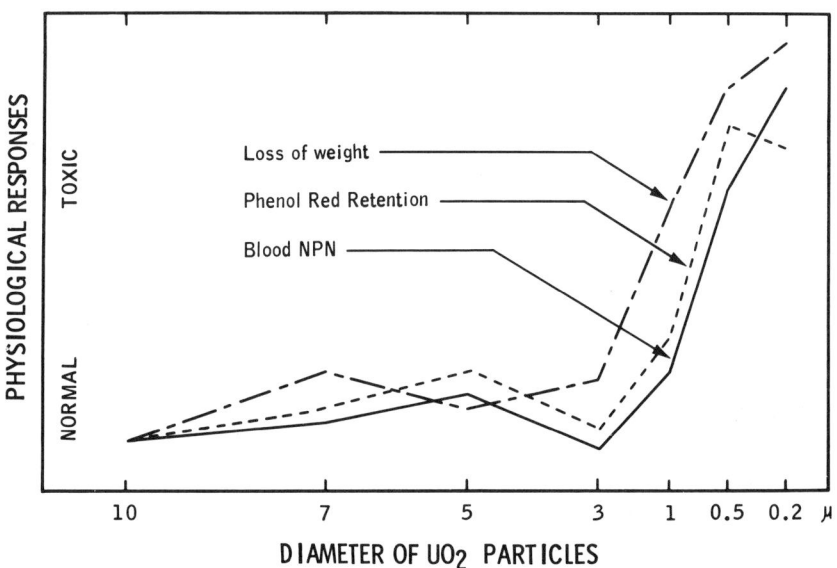

FIG. 2-16. Effect of particle size on physiologic activity (from Dygert et al., 1949).

Except for physical blockage of airways, particulates greater than 10 μm in diameter are largely ignored in physiologic and public health considerations. Only 1–2% of the total dusts, the small particulates, accounts for most of the damage. Casarett and Doull (1975) review the chemical toxicity of some inhaled metals. Practical examples are given in Table 2-7. Specific information about the inhalation toxicity of metals is summarized in Volume 2.

Translation of Inhalation Data

In comparative toxicology it is desirable to relate data from inhalation studies to those of other modes of administration that use milligrams of compound per kilogram body weight with specified times of dosage and observation. The complexity of inhalation makes evaluation of all components difficult. How much material was inspired? How much was deposited? What is the mix of pollutants? What is the particle size? What is the solubility? How much was expired? Although inhalation problems are more readily discernible than equivalent problems in oral administration, they are neither different in concept nor greater in magnitude for experimental design or theoretical treatment.

In both systems, highly soluble material will be readily absorbed during the initial phase of tissue saturation, and insoluble particles and complexes will be poorly absorbed. Inhaled material that is deposited in the alimentary tract provides the complexities of both modes of administration. The material so deposited is less reactive, however, because the lungs are considerably more efficient for absorption of toxic metals than is the intestinal tract. Thus, inhaled material that is processed via the alimentary tract usually has a minor impact in the toxicity of metals. This complexity is not enough, however, to prevent drawing parallels between inhalation and ingestion. They are comparable in the following ways:

1) The condition of the mammal modifies rates of passage through the absorption area. The half-life of rapidly removed material deposited in the total respiratory tract approximates 1 day (Boyd, 1972b), while that in the digestive tract is 1 to 2 days. Irritants may speed the former by a factor of 2, and the latter by a factor of 20.
2) The total quantity and the physical and chemical state of the metal is important in determining how much will be absorbed. If the material is highly soluble, all may be absorbed; previous discussion has indicated that the alveolar volume (70% of tidal volume) is acceptable for estimates of the quantity inhaled.

TABLE 2-7. Summary of Metal-Induced Respiratory Diseases Called Pneumoconiosis[a]

Dust type	Metals	Tissue	Response	Designation
Quartz	Si and others	Lung parenchyma Lymph nodes Hilus	Collagen, with some reticular nodes	Silicosis
Coal or graphite	Si and others	Lung parenchyma Lymph nodes Hilus	Primarily collagen nodules	Anthracosilicosis or graphite lung
Kaolin	Al and Si $[Al_2(SiO_2)_5)OH_4]$	Lung parenchyma Lymph nodes Hilus	Primarily collagen nodules	Kaolinosis
Diatoms	Si	Lung parenchyma	Diffuse fibrosis	Diatom lung
Talc	Mg and Si $[Mg_6(SiO_2)OH_4]$	Lung parenchyma Lymph nodes	Collagen nodes Granulomata	Pleural sclerosis
Asbestos	Fibrous silicates (with others)	Lung parenchyma	Diffuse fibrosis	Asbestosis

[a] Adapted from Casarett and Doull (1975).

3) The chemical form of associated molecules is very important in determining the availability of the metal for absorption. Complex mixtures have been studied much more in oral than in inhalation experiments.
4) Some of the material administered may be excreted with little or no processing. This finding has been much discussed in digestion; comparable processes occur in inhalation:
   a) Aerosols less than 0.1 μm in diameter have so little mass that they remain suspended during respiration, and, unless readily soluble in the mucoid coating of the pulmonary tract, they are expired.
   b) Particles greater than 10 μm in diameter are largely stopped in the upper respiratory tract and swept into the alimentary tract with no absorption in the lungs.

   Concern for particle size is a refinement of inhalation toxicology that has occupied oral toxicologists less than it deserves; note the different effects of particulates, colloids, and radiocolloids as discussed in Chapter 1.
5) Particulates may be carried by phagocytes or by lymph, or they may be dissolved and carried in body fluids to a variety of target organs from both systems.
6) Excess quantities are treated in a manner that decreases absorption. Within physiologic limits, the lungs will have increased fluid secretion and cilia motion, and may form tubercles to wall off deposits of the excess. The alimentary tract typically responds with faster elimination.

This outline of similarities allows an approximation of comparative toxicity to be made either on the simple basis of milligrams of dose per kilogram body weight or on the basis of a molal concentration. Inhalation doses are usually given as milligrams of material per cubic meter for a stated time. The standardized minute-volume of respired air for a number of species (Table 2-1) was calculated from the data of Gleysteen and Stroud (1974). This standardization allows a simple calculation for an approximation of the total material inspired.

Two concepts must be understood. First is the rate of distribution of a nonabsorbed gas in the alveoli. Eger (1974) states that normal breathing will establish alveolar concentration to be 98% of that of a constantly inspired gas within 2 min, provided none was absorbed. Second is the tremendous absorption of an extremely soluble gas during the first minutes of inhalation before any significant equilibrium with the tissue reservoirs has been established. Although few data have been found to establish either concept,

MODES OF INTAKE AND ABSORPTION

the information in Fig. 2-17 illustrates both. Sechzer (1963) found that mathematical models agree reasonably well with experimental data.

The rates at which different concentrations of gases will equilibrate with lung fluids and tissues depends on the solubility of those gases in the blood and their concentrations in inspired air during normal respiration (Fig. 2-17). Poorly soluble helium equilibrates within 2 min. Low concentrations of a soluble material, as modeled by 1% acetone (experimental data are not available), show virtually no equilibrium within 10 min. These models of (1) helium distribution and (2) acetone absorption in the lungs indicate that usual concentrations of a soluble gas would be completely absorbed during the first 2 min, while equilibrium with body tissues is negligible. A small quantity of soluble gas exhaled would be dissolved in the water droplets exhaled. At 37°C, saturated vapor provides 44 mg water/liter. Association of the soluble gas with this small amount of vapor would be minimal when compared with the liquid covering the 20-m² surface of the human lungs. That, plus a blood flow time of 1–2 min through the body, assures that the initial absorption rate will continue for several minutes before an effective equilibrium between alveolar air and blood can be detected. Equilibria between alveolar air and blood and blood- and vessel-rich organs (brain, spleen, liver, kidney, and the endocrine glands), blood and muscle and skin, blood and fat, and blood- and vessel-poor organs (bone

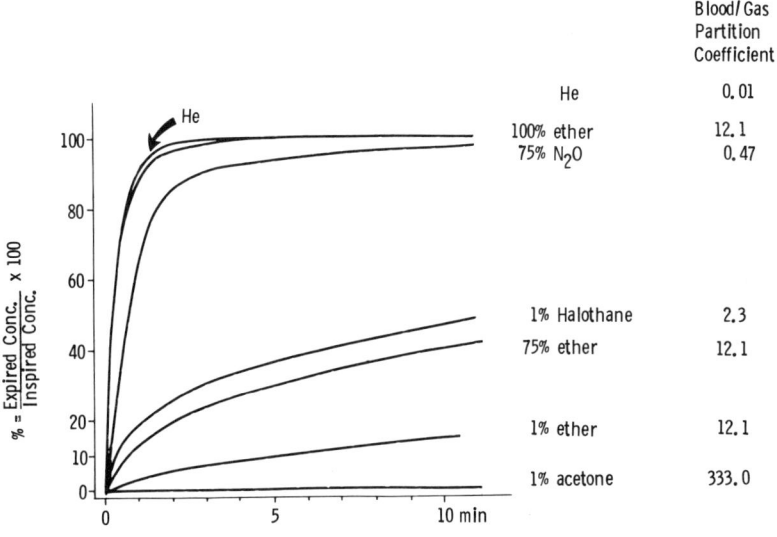

FIG. 2-17. Equilibria between inspired and expired gases at different absorption times.

and connective tissues) are established at 1 min, 2–5 min, 1–1 ½ hr, ~ 3 hr, and > 3 hr, respectively (Eger, 1974).

The foregoing discussion suggests that low doses of toxic compounds which enter the alveoli will be absorbed according to their physical and chemical characteristics. To the limits of blood and tissue saturation, this relationship is comparable with the fate of ingested materials that are absorbed at different rates, depending on the solubility of the material, the quantity given, the size and chemical character of associated particles, and the condition of the organism. Insoluble material is poorly absorbed in both the lungs and the alimentary tract.

Although supporting data are nebulous, this argument indicates that information from inhalation toxicity (milligrams of substance per cubic meter of air and time of inhalation) can be incorporated into dose–response data banks using the classic milligrams-per-kilogram measurement. Standard values useful in this conversion are given in Table 2-1. The simple formulas needed are:

$$\text{amount/vol} \times \text{vol/min} = \text{amount/min}$$

$$\frac{\text{amount/min} \times \text{min}}{\text{kg}} = \text{amount/kg}$$

This information is utilized to incorporate inhalation data into concepts of comparative toxicology (Volume 2). Molal concentrations of inhaled metals will be used to calculate pT values for comparative toxicology:

$$\text{Molar concentration} = I \times MW^{-1} \times kg^{-1}$$

where I is the mg intake and MW is the molecular weight,

$$I = CAT$$

where C is the concentration in mg/m^3 under the stated conditions, A is the air respired in m^3/min, and T is the time of exposure in minutes. These formulas provide ready translation of the usual inhalation data into milligrams-per-kilogram dose, molal concentration, or pT. Since respiration is rarely measured, standard values may be used (Table 2-1).

Guidelines for inhalation toxicology studies are provided by the Canadian Food and Drug Protectorate (1965).

DERMAL ABSORPTION AND ADMINISTRATION

The skin is the elastic, rugged, and complex exterior organ covering the body. In adult humans the area of the skin is approximately 1.8 m^2.

Under normal, uninjured conditions, the skin prevents the absorption of toxic material—especially inorganic electrolytes, particulates, molecular aggregates, and high-molecular-weight substances—by virtue of its integrity, its water-repellent nature, its unique structure, its regenerating capacity, and its metabolism (Rothman, 1965; Moschella *et al.*, 1975). Unfortunately, the skin did not evolve to prevent the absorption of lipid-soluble metallic pollutants. Animals with a thick cover of body fat retain these metallic toxicants in this fat reservoir; this retention prevents the access of such toxicants to sensitive tissues. Casarett and Doull (1975) discuss briefly the comparative aspects of absorption.

The skin has well-established homeostatic mechanisms to prevent loss of essential minerals and to excrete excess of some minerals in sweat, sloughed cells, and hair. The use of subcutaneous solutions, suspensions, pellets, and discs is becoming commonplace for chronic therapy or preventive medicine. Incorporation of vasodilators or vasoconstrictors in these medications provides further control of the absorption rate. While larger volumes of isotonic fluids are readily absorbed following subcutaneous administration, irritating solutions are more painful by this mode than by intravenous or intramuscular administration (Meyers *et al.*, 1972). Serious tissue reaction to material applied subcutaneously or interdermally may cause sloughing, irritation, or encapsulation. Sloughing and irritation may be observed following inunction or topical application to the skin surface. These local reactions complicate the evaluation of overall toxicity (Weil and Rostenberg, 1969). Skin sensitivity and dermatitis are encountered with mercury, nickel, thallium, and a few other metals.

The major components of the skin (Fig. 2-18) are the epidermis, dermis, and appendages. Elastic fibers, fascia, nerves, and blood and lymph vessels connect the skin to subcutaneous tissues. The epidermis is composed of stratified squamous epithelial cells and is thickest on the palms and the soles. The epidermis consists of five layers or strata—corneum, lucidium, granulosum, mucosum, and germinativum—which are sometimes identified by other names. The stratum corneum is composed of keratinized cells and is horny in nature; it is a physical barrier to light, heat, microorganisms, and most chemicals. Melanin pigment is present in the stratum germinativum, the major site of mitotic division.

The dermis, or corium, is the true skin lying directly beneath the epidermis; it consists of white collagenous and yellow elastic fibers. Nerves, blood and lymph vessels, hair follicles, sebaceous glands, and sweat glands are embedded in the dermis. The dermis is organized into the papillary segment close to the epidermis and a reticular portion between the papillary layer and subcutaneous tissue.

Appendages associated with the skin are hair, nails, and sebaceous and

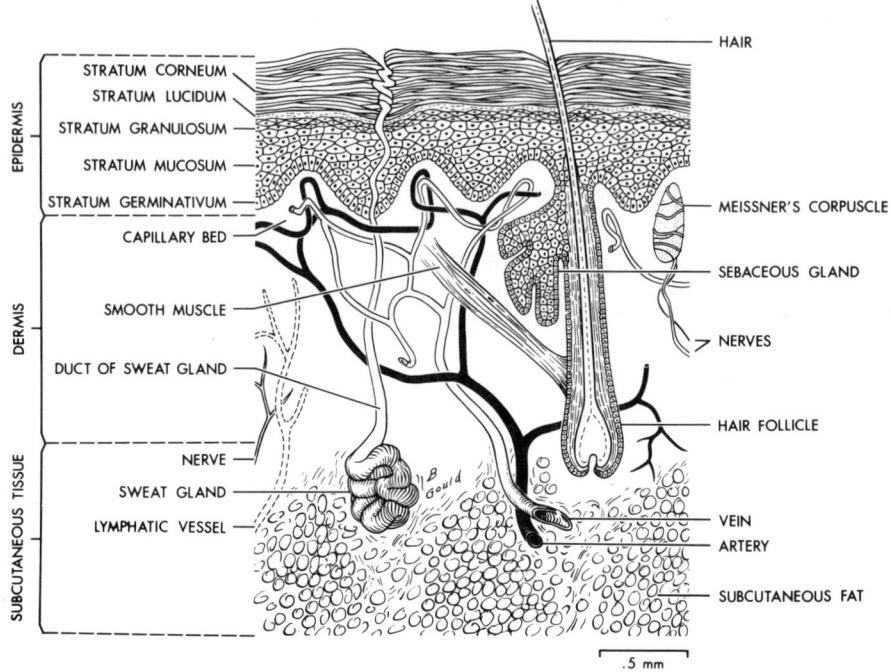

FIG. 2-18. Diagram of skin structures.

sweat glands. Hair appears on almost the entire body and affords some protection against environmental toxicants. Hoofs and nails, modifications of horny epidermis, are composed of hard keratin. The nail plate, arising from the proximal nail fold and attached to the nail bed, grows about 1 mm per week in healthy humans. Hair follicles, which include the root, are embedded in the skin, while the visible portion is the shaft. The epithelial cells of hair follicles undergo mitotic division, and the new cells push upward, keratinize, and form the horny layer of the shaft as it lengthens about 1 mm per day. When the skin is involved in the excretion of heavy metals, such as mercury or arsenic, the metal is attached to the cysteine of keratin of the hair shaft. Hence, hair and nails serve as indices for the presence and excretion of toxic metals. The analysis of hair for minerals is presumptive evidence for metal intake, although this method is more useful for arsenic—it can indicate excess intake—than for microquantities of environmental metals. It is difficult to determine whether minute quantities of metals in hair were adsorbed from exogenous sources or incorporated from endogenous supplies. Washing in water, alcohol, or detergent provides suggestive evidence, but these treatments cannot release strong adsorptive bonds, including chelation, which may occur with exogenous metal. Lockeretz (1973) indicates that errors in hair analyses from adsorbed metals do not necessarily reflect the metabolic overload of that metal.

The normal intact skin is usually impermeable to water, carbohydrates, and protein; gases and volatile substances can pass through the epidermis. The numerous follicular and sebaceous gland orifices can serve as channels for absorption of fat-soluble toxicants.

Sebaceous glands arise from the walls of hair follicles and produce oily sebum, the lubricant of the skin surface. The secretory portion of the exocrine sweat gland may be located in subcutaneous tissue below the dermis, and the excretory duct spirals through the dermis and epidermis toward the surface. Each duct is lined with secretory epithelium continuous with the epidermis. Sebaceous glands, which contribute sebum, are associated with hair follicles.

The microflora may contribute more living cells to skin than does the host; there are up to 10^{10} bacteria per square centimeter of human skin. The predominant species is *Staphylococcus epidemidus,* along with a considerable number of anaerobic *Propionobacterium acne.* Successive applications of sterile adhesive tape to one area show that bacteria penetrate to the capillary bed (Ulrich, 1965). Excessive humidity may dramatically change the indigenous flora, especially when a bandage or other barrier excludes air from the skin surface. Marples (1974) found that the groin and the axilla harbor the most bacteria. Low-molecular-weight fatty acids, produced by anaerobic bacteria, contribute greatly to the exclusion of many other microbic species from the skin habitat. This acid mantle serves as an antibacterial agent (Maibach *et al.,* 1973).

Accidental or intentional contact of the toxicant with the skin is the most common exposure of man and animals to metallic toxicants (Wilkinson, 1965). Percutaneous absorption of chemical toxicants is the transfer of the toxicant from the outer surface of the skin through the horny layer of the epidermis and into the systemic circulation. The pathway is partly transfollicular via the permeable cells of sebaceous glands and the follicular wall, and partly through the epidermis. The extent of absorption depends on the physicochemical properties of the toxicant, such as molecular size, water and lipid solubility, extent of ionization, and hydrolysis of the toxicant at the pH in the epidermis and dermis. Local factors, such as temperature and blood circulation to the skin, also influence the rate of absorption. Whereas the small size and lipid solubility of the toxicant molecules enhance the rate of percutaneous absorption, the different layers of the skin present physical barriers to absorption. The bottom half of the stratum corneum and the stratum lucidum are extremely dense and are composed of keratin material containing less than 10% water. This layer appears to be electronegatively charged; it repels anions and prevents cations from penetrating deeply and is the most effective barrier of the skin. The properties of the skin as a barrier vary considerably among species. The skin of the axilla, groin, and head absorb much more readily than does the skin of the back or thigh. Man has a thicker skin than other mammals. Pig skin has a higher diffusion rate for water and for water-soluble substances than rat or guinea pig skin (McCreesh, 1965). Application of lipid solvents and detergents to the skin may remove the protective fat layer or provide water retention of the skin and increase permeability to metal ions. Bettley (1965) found that lauryl salts would readily penetrate the skin. Breaks in the epidermal continuity or cutaneous injury may enhance percutaneous absorption of metallic toxicants. Shaving probably enhances skin absorption. Keller (1958) found that shaved skin of mice would allow ready penetration of virus, when the virus was gently placed on a healthy skin surface. This finding suggests that minute quantities of any lipid-soluble material would be rapidly absorbed.

The cornified epithelial cells form an effective barrier against water-soluble intoxicants. These dead cells permit no active transport, and the slow diffusion is inversely proportional to molecular weight (Marzulli *et al.,* 1965). The constant shedding of these cells tends to release any adsorbed toxicant from prolonged contact with the skin. The flow of nutrients by seepage from the capillary bed, sebum through hair follicles, and sweat through sweat ducts provides other mechanisms that supplement the shedding of skin and hair. These movements may effectively retard a slowly diffusing metallic toxicant from passing through the skin. The natural process of drying to stimulate keratinization, cornification, and shedding is a homeostatic mechanism of the skin that decreases the entry of metallic toxicants into the body. Sweat is an effective excretory system for heavy metals such as copper, cadmium, zinc, and lead (Arena, 1974; Suzuki, 1975).

Nonpolar and lipid-soluble compounds may readily penetrate the fatty, horny layer of the epidermis. Metallic toxicants such as mercury salts (both inorganic and organic) and other metallic toxicants that dissolve in or distribute in the lipid medium are absorbed readily. Absorption or tissue reactions have been noted for the following metal salts when administered by inunction: copper, gold, beryllium, zinc, mercury, thallium, arsenic, chromium, cobalt, and nickel. Special compounds such as lead tetraethyl are easily absorbed. The placement of these metals in the periodic chart suggests that other metals should be suspected of having similar properties (i.e., cadmium, silver, or tetramethyl tin).

The increased absorption of metallic toxicants applied to the skin and covered by occlusive bandages is due to a variety of factors. In addition to the prolonged time of contact with skin and negligible loss from the site of application, the high humidity conditions under the cover increase metal absorption. The cover allows a changed arena for action; the secretions of the skin and the low-molecular-weight fatty acids produced by the bacteria solubilize some insoluble metal compounds to facilitate greater absorption. Since covered keratin retains water, cornification cannot proceed to form a hardened epithelial layer, and water-soluble material diffuses much more readily through the moist keratin-containing cells than through the dry lipoprotein remains of the keratinized cells. Miedler and Forbes (1968) reviewed cutaneous mercury poisoning from covered inunctions. The presence of dimethyl sulfoxide (DMSO) or other dimethyl derivatives greatly increases the absorption of both lipid- and water-soluble substances. Keratin denaturization may occur. The water-soluble metal toxicants, following their passage through the acid mantle of the skin, may precipitate at the changed pH conditions in the dermis. The presence of solid foreign matter will initiate a spontaneous dermal reaction; the macrophages, leukocytes, and mast cells will infiltrate and wall off the foreign material. Phagocytosis ensues, and the particulate toxicants are removed from the dermis. Granuloma of these cells results when phagocytosis is slow or unsuccessful. Silicon, beryllium, and zirconium granulomas have been described (Jarrett et al., 1961).

While the skin surface is the major entry route, the openings of the dermal appendages, hair, and sebaceous glands permit the entry of toxicants. The thick and densely packed barrier described earlier terminates at the level where the sebaceous gland enters the follicle. Toxicants can diffuse through the continuum of the liquid phase of the skin to sebaceous glands. The sebum-coated hair follicle provides a ready pathway for lipid-soluble metallic toxicants to diffuse from the skin surface into the dermis and the capillary bed without going through the outer layer of epidermis. The capillary network of the dermis, the lymphatic system, and the macrophages will quickly clear or transport the toxicants to other tissues. Since the lymphatic vessels have no musculature, gentle massage of the skin will greatly increase the rate at which these toxicants reach the reticuloendothelial system. The sebaceous glands and associated vascular systems, with their extensive blood supply, are permeable to toxicants. Sweat glands do not appear to provide a significant portal of entry.

Experimental studies on cutaneous absorption include three modes of administration: inunction and intraepidermal or intradermal injection. Inunction is the topical application to the surface of the skin, which may be covered or left exposed. Different anatomic regions vary considerably in their absorption capacity. Parish and Champion (1973) indicate the relative absorption of hydrocortisone in man: forearm, 1; back, 1.7; scalp, 3.5; forehead, 6; underjaw, 13; scrotum, 42. Inorganic compounds have been less well studied. Since hair and protected areas decrease water evaporation, the perineum and armpits would be expected to provide greater absorption of soluble metal salts than would be found on the thigh, the back, or other bare areas. Toxicants could be injected into either the epidermis or the dermis. Compounds injected into the dermis will effectively bypass the lipoidal barrier of the epidermis. Water-soluble materials may diffuse into lymphatic vessels and the capillary bed of the dermis. Microparticulate material would be expected to be phagocytized slowly in the deep epidermal layers and relatively quickly in the dermis. Subcutaneous injection of metallic toxicants will result in the absorption of these toxicants in a manner similar to that of intramuscularly injected compounds.

Dermal administration in metal toxicity is limited to a relatively few studies using subcutaneous injection. Reviews on the toxicity of dermally applied drugs and toiletries (Rostenberg and Coulston, 1965) contain little on metal toxicity. Goldstein *et al.* (1974) indicate that some drugs are absorbed as rapidly from subcutaneous tissues as from intramuscular injections. The rate of absorption can be controlled by appropriate diluents and by varying the form and area of the material injected.

Hair and reasonable avoidance of close trappings (harness connectors) prevent contact dermatitis in most mammals, while accouterments of clothing and jewelry offer continuous contact for metal toxicity to develop in man. Some metals are primary agents in allergy and dermatitis, and a few act as sensitizing agents. Prolonged contact and increased concentration of metals produce increased incidence and severity of dermatitis and increased sensitization reactions. Sweat or tight coverings accentuate contact dermatitis by increasing local humidity and permitting greater direct interaction between host cells and agent. Aqueous solutions of many metal salts are so poorly absorbed that they may be useless in the detection of metal sensitivity. Covering the skin changes these characteristics and increases the absorption of metals.

Prophylaxis in metal dermatitis is frequent cleansing, avoidance of tight clothing, reduction of sweating, and avoidance of continued contact of metals with the skin. Treatment of mild dermatitis includes topical corticosteroid creams, sprays, and lotions (Domonkos, 1971).

OTHER PARENTERAL ROUTES

Metal toxicity is usually experienced by eating, inhalation, or dermal contact. Parenteral routes are less usual and, except in vaginal application, are of little direct public health concern. Subcutaneous (s.c.), intravenous (i.v.), intraperitoneal (i.p.), and intramuscular (i.m.) administration provide the desired quantitative data needed for comparative toxicology. Data from oral administration can be correlated with data from parenteral administration only when the amount of absorption and the changes during digestion are evaluated. The quantitation of data from inhalation is not comparable with that of any other mode of administration, because the amount absorbed is rarely determined.

Intravenous Administration

Intravenous, intraarterial, or introcardiac administration is the introduction of material by injection, whether immediate or by continuous reservoir, into the veins, arteries, or heart, respectively. Intravenous administration is the most commonly used of these most direct avenues to tissues, and provides exact dosage with no concern about digestive changes or absorption. Slowly administered material is continuously diluted with fresh blood; the patient may have extraneous reactions to this mode of administration, called intravenous infusion. Complications may include trauma to small mammals, thrombus, embolism, infection, and pyrogenic and anaphylactic reactions. It is therefore one of the most hazardous methods of administration. Intraarterial administration may be desired for selected studies with a specific organ, but has been little utilized for metal toxicity studies.

Following injection into the vein, the metal compounds behave differently according to their physicochemical properties. High doses injected rapidly into the vein will produce sudden colloidoclastic shock. Soluble metal salts, which remain soluble at the pH of blood and which do not alter the osmotic equilibrium of the blood, are transiently bound to plasma proteins, amino acids, or metabolites such as citrate, bicarbonate, phosphate, and ATP, and are rapidly cleared from the blood and transferred to other tissues. Some metal ions, such as those of mercury or copper, bind to erythrocyte cells and are sequestered in the erythrocytes. Insoluble metal salt suspensions or salts of Group III metals, which are hydrolyzed at the pH of the blood into their insoluble or colloidal hydroxides or oxides following injection, are trapped in the blood by leukocytes; they are also trapped by other cells of the reticuloendothelial system or hepatic cells. Metal compounds of this type are cleared slowly from the blood and transferred to soft tissue and bone.

Intravenous administration is good for exact and reliable quantitative data on acute toxicity of metals and for short-term studies; it is less suitable for chronic toxicity studies, and is not used for life-term studies involving two or three generations.

Intraperitoneal Administration

Intraperitoneal injection is the administration of chemicals into the peritoneal cavity (abdominal fluid) of animals for evaluation of their biologic action and toxicity. This type of injection is done in humans only on rare occasions, and under special conditions in neonates, but it is valuable for treating animals clinically.

The peritoneum is the serous or mesothelial membrane lining the abdominal cavity. The outermost covering of the gastrointestinal tract within the abdominal cavity is called the visceral peritoneum, or serosa, and the parietal part lines the abdominal musculature. The peritoneal cavity exists in many areas of the abdomen between the visceral and parietal peritoneum. The peritoneum acts as a semipermeable membrane; it is also connected with an efficient vascular system. The peritoneal cavity contains a variable amount of fluid with a pH ranging from 5 to 7.

Soluble metal compounds, which remain stable and soluble at the pH of the peritoneal fluid, are rapidly absorbed across the visceral peritoneum and are transferred to the liver via the portal circulation. Hepatic cells process absorbed material before it reaches other tissues. If the injected metal is particulate or hydrolyzed into particulate form at the peritoneal pH (e.g., salts of Group III metals), these microparticulates are phagocytized by the invading macrophages and scavenged into the reticuloendothelial system, which includes the liver and spleen. The accumulation or converging of macrophages is the primary response of the body to the introduction of toxicants into the peritoneal cavity. Thus, the intraperitoneally injected metal is subjected to the special detoxication or metabolic transformation mechanisms existent in phagocytes and the liver, with possible excretion in the bile, before it can reach other tissues. The higher LD_{50} values for metal toxicants in rats administered by intraperitoneal injection over those administered by subcutaneous or intramuscular injections support this view. Thus, although the absorption of soluble metal salts from intraperitoneal injections is rapid, the toxicity is less, due to possible detoxication by the liver.

Intramuscular Administration

Intramuscular administration is the injection of soluble or suspended material deep into the larger muscles for quick absorption into the blood. The absorption or clearance from injection sites depends on the lipid or aqueous solubility and stability of the injected metal compounds in the internal environment of the muscle. The ideal site of injection is deep within the muscle and away from major nerves and blood vessels; the best sites are the thigh and gluteal muscles. This may be the best method for administration of irritant solutions.

Toxic metallic compounds, that are stable and remain soluble at the pH of the muscle without chemically reacting with active components of muscle, will be cleared rapidly from the site of injection into the blood.

Some metal compounds, such as silver salts, will react with sulfur-containing amino acids or proteins and will remain immobilized permanently at the site of injection. Compounds that separate as insoluble material due to hydrolysis or other reactions will be slowly removed by phagocytosis into the reticuloendothelial system. The absorption of metallic toxicants from the sites of intramuscular injection depends on the physicochemical characteristics of the toxicants. Absorption into the blood is rapid for neutral, aqueous solutions and nonpolar compounds, and should be complete within a short time. The rate of absorption can be influenced by suspension of the material in a suitable medium, such as oil.

Vaginal Administration

Vaginal administration is important because it provides a readily accessible route for foreign material to be presented directly to an internal organ without breaking any anatomic barrier. The internal membrane lining permits efficient absorption. A variety of toxic actions are reported from clinical material, and the use of copper as a vaginal insert presents a variety of problems. Metal compounds, such as phenylmercuric acetate used in some contraceptive creams or jellies, may cause difficulties.

Other Routes

Other methods of parenteral administration include ocular, auricular, testicular, intrathecal, urethral, and bladder applications, and injections into bone marrow, arachnoid cavity, and brain. Intrathecally injected metal compounds bypass the blood–brain barrier, which permits the exposure of the brain to high levels that could not be achieved by any other route of administration. Toxicity following intrathecal administration of metallic toxicants varies in comparison with other modes of administration, and depends on the primary target tissue of the toxicant and on the barriers to translocation of the toxicant from the site of administration. Experimental injection into the fetus constitutes another facet of metal toxicity in mammals. Most of the routes mentioned above assess the toxicity of the metallic

TABLE 2-8. Comparison of Acute Toxicity of Organic Sulfur Compounds by Different Modes of Administration[a]

Compound	IP LD$_{50}$ (mg/kg): Rats	Oral LD$_{50}$ (mg/kg): Rats	Inhalation LD$_{50}$ mg/m^3 Mice	Inhalation LD$_{50}$ mg/m^3 Rats	Eye irritation: Rabbits	Skin LD$_{50}$ (mg/kg) Mice	Skin LD$_{50}$ (mg/kg) Rats	Skin LD$_{50}$ (mg/kg) Rabbits
Ethanethiol CH$_3$CH$_2$SH	226	682	2,770	4,420	slight			
Propanethiol CH$_3$(CH$_2$)$_2$SH	515	1790	4,010	7300	moderate			
2-Methyl-1-propanethiol (CH$_3$)$_2$CHCH$_2$SH	917	7168	>25,000	>25,000	very slight			
2-Methyl-2-propanethiol (CH$_3$)$_2$CSHCH$_3$	590	4729	16,500	22,200	slight			
Butanethiol CH$_3$(CH$_2$)$_3$SH	399	1500	2,400	4,020	slight			
Hexanethiol CH$_3$(CH$_2$)$_5$SH	396	1254	528	1,080	none			
Methyl heptanethiol C$_8$H$_{17}$SH	12.9	83.5	47	51	slight	19.1	1594	600
Benzenethiol C$_6$H$_5$SH	9.8	46.2	28	33	severe	6.5	300	134
α-Toluenethiol C$_6$H$_5$CH$_2$SH	373	493	178	>235	slight			

[a]From Fairchild and Stockinger (1958).

toxicant on a special target tissue and do not involve absorption or translocation. Ophthalmic application of toxins was reviewed by Beckley (1965).

SUMMARY OF MODES

A comparison of the usual modes of intake is provided in Table 2-8. This compilation by Fairchild and Stockinger (1958) indicates that intraperitoneal injection of sulfur compounds gives about 5 times greater acute toxicity than does oral intake. Comparable data are now available for metals. The oral modes of administration are much more effective than inunction. The skin is an effective barrier to most environmental hazards. Unfortunately, ocular data are not directly comparable with those from other modes of administration. There is fair correlation of activities from different compounds between inhalation and oral intake. Dautreband (1962) and Patterson (1965) indicate that of the lead deposited as lung particulates, about 70% was absorbed; about 10% was absorbed from similar particulates given by gavage. Cember et al. (1956) found that particulates of barium sulfate were removed from the lungs within a week, while virtually no barium was absorbed when barium sulfate was given orally. Toxicity is usually greater when the metal is administered parenterally than when it is ingested (Boyd, 1972a). Intravenous administration of soluble metallic toxicants in a solution results in rapid systemic distribution of the toxicant to all organs of the experimental animal; the rapidity is limited only by the time required for blood circulation and for translocation of the toxicant from the capillaries to the extracellular fluid. Oral intake, inhalation, and dermal contact are the more common modes of intake that are generally met with in accidental or intentional poisoning. Inhalation results in more systemic distribution of the toxicants than is attained with either oral intake or dermal contact.

This review of the various routes of administration of toxicants shows different degrees of absorption and consequently different degrees of toxicity for the same toxicant. Less well studied are the changes in absorption, retention, and toxicity of metals under conditions of stress or impaired metabolic function. The toxicity of metals may be greater in unhealthy than in normal persons.

3

DETOXICATION, EXCRETION, AND PHYSIOLOGIC HOMEOSTASIS

Animals develop both specific and nonspecific defenses following exposure to toxic materials, including metals and their compounds as well as natural toxins and organic toxicants. Psychic instinct and past painful reactions to these toxicants guide the animals to avoid reexposure to some toxic materials. Following oral ingestion, animals may expel toxicants by expectoration, vomiting, or purgation, which are primary physiologic responses to toxicants. If toxicants gain entry into their tissues and cells, animals attempt to detoxify these life-endangering substances by gene-controlled, special metabolic processes. Detoxication is the biochemical process of transforming a harmful substance into a less harmful substance. Such special metabolic processes are dormant in all species of animals and are activated immediately on exposure. When animals are exposed to a toxicant for the first time, the metabolic processes of detoxication are induced. If the chemical and biologic properties of the toxicants are similar to those of an essential metal, homeostatic mechanisms for essential nutrients are utilized in detoxication.

Detoxication mechanisms include: (1) biologic oxidation, reduction, or hydrolysis of the metal toxicants into less toxic forms, e.g., hydrolysis of salts of Group III metals in the gastrointestinal tract of animals; (2) incorporation of the toxicants into macromolecules that are harmless, e.g., cadmium and mercury in metallothionein; and (3) sequestering of the toxicants into less toxic forms inside special structures or organelles, which deprives them of access to the target tissues, e.g., colloidal forms of toxic insoluble metal salts in lysosomes, and immobilization of lead and lanthanide metals in bone. The first of these biologic changes tends to transform toxicants into products that are less lipid-soluble, more water-soluble, and more acidic than the original toxicant, to enhance excretion in the urine. Excretion of toxic or detoxified compounds occurs mainly through urine or feces or both; exhaled air, sweat, and tears serve as minor excretory media for metals. References for this general information may be found in texts on pharmacology, physiology, and nutrition.

DETOXICATION

Detoxication is the secondary physiologic response of the animal to a toxicant. Under conditions of stress imposed by the toxicant, the animal mobilizes biochemical mechanisms that usually convert the toxicant into a less harmful and an easily excretable form. Detoxifying systems are effective against chronic toxicity involving small but repeated doses of a toxicant. Increased levels of this toxicant in the animal exceed the capacity of the detoxifying systems, and toxicity symptoms become pronounced; they can include death. Detoxication is less effective against acute and sudden exposure to a single large dose of a toxicant, especially toxic metal compounds. The animal may succumb to the lethal effect of a single dose of a given toxicant that is much less than the nonlethal cumulative dose of the same toxicant taken chronically. Detoxication mechanisms are not available in mammals for all toxic metals; those for some of the metals are complex and cover a broad spectrum of physiologic and biochemical reactions.

Prior exposure to and transient traumatic experience of the toxicant guide the animals to avoid a second exposure. Laboratory experimental animals, especially monkeys, tend by instinct to avoid eating special diets that contain "odorless" toxic metals. Horses escape ingested lead poisoning because few live near smelters, and their behavior patterns do not include licking old paint cans or eating peeling paint. Humans avoid inhaling toxic vapor and move away from the source when the toxicant irritates the mucous membrane of the eye, nose, and respiratory tract. All these actions can be grouped under psychic reactions to avoid exposure to toxicants.

Following exposure to toxic metals, the animal responds with a number of physiologic detoxication procedures, some of which are: (1) inhibition of absorption of the toxicant from the gastrointestinal tract and from the sites of parenteral injection (excluding the intravenous route); (2) removal from the blood of the absorbed metal ions by efficient transport to other tissues that can detoxify the toxicants, or by sequestering in the erythrocytes; (3) removal of toxic and insoluble metal compounds from the blood and body by phagocytosis into circulating macrophages, and transfer of the compounds from the macrophages into the reticuloendothelial system for penultimate excretion into the digestive tract and final excretion in the feces; (4) solubilization of the metal complex by oxidation, hydrolysis, chelation, or other complexation to provide increased efficiency of urinary excretion; and (5) rendering the toxic metal or ion less harmful within the cell by combination with nonessential metabolites, thereby sparing the essential metabolites, or by incorporation into existing or new proteins; other nontoxic metal ions can interact with the toxic metal, rendering it less harmful. A brief explanation of each of these detoxication procedures follows, and they will be referred to repeatedly in Volume 2.

The near-neutral pH of digesta in the intestine ensures hydrolysis of the soluble metal salts of the metals from Group III, zirconium and hafnium from Group IV, and bismuth from Group V. The resulting metal hydroxides, oxides, or oxysalts are insoluble and are negligibly absorbed from the digestive tract. Poor absorption renders these metals less harmful and is considered a detoxication process. Osmium salts react with the lipids of the mucosal wall and are retained there permanently. Argyria at the site of injection, following the parenteral administration of soluble and toxic silver salts, appears to be a specific detoxication mechanism for silver. The formation of black silver sulfide due to reaction with sulfur-containing proteins and amino acids provides permanent retention at the site of injection. This silver compound is not mobile and is unavailable for any other toxic action. Bismuth salts and silica are also retained permanently at the site of parenteral injection. Other metals are phagocytized and transported throughout the body by macrophages; these cells also transport particles from the lung into the alimentary tract or the reticuloendothelial excretory system or both.

Following absorption into the blood, metal ions may be rendered less toxic by a variety of detoxication procedures. Clear examples of toxicants being made more harmful are rare for metals. Lanthanide cations and similar ions react with serum phosphates or with chelating agents such as lactate and citrate and are rapidly removed to tissues, such as bone, where they are immobilized in a harmless, sequestered condition. Lead is immobilized in a similar way. Therapeutic administration of synthetic chelating agents, such as EDTA, allows the lanthanides and the actinides to form stable and soluble chelates that are excreted rapidly in the urine. Toxic mercuric ions are sequestered in erythrocytes and immobilized temporarily. Other toxic ions complex with serum proteins, such as transferrin, and are rapidly removed from the blood. Germanium ions can be oxidized into more soluble germanates for efficient renal excretion. Lanthanides and metal salts of Group V at higher concentrations separate as colloidal or particulate forms in the blood at the neutral pH. The formation of less harmful colloidal forms of the minerals from the toxic ionic forms is detoxication. These forms are slowly removed from the blood by phagocytosis and transported to the reticuloendothelial system. During this sequence, the toxic ions are temporarily kept in a harmless condition. The toxic materials is sequestered in macrophage lysosomes, which are transformed into secondary lysosomes or vacuoles. In the case of some carcinogenic metals, it is assumed that prolonged stay in the lysosomes activates some cellular constituent to become carcinogenic.

Some toxic minerals stay sequestered in the reticuloendothelial system before final excretion through the bile into the intestine. Some metals such as indium are activated to become more toxic when trapped in the reticuloendothelial system. This activation can be prevented and indium ions detoxicated by saturating the reticuloendothelial system with the harmless compounds of Group III metals, such as aluminum hydroxide, thus preventing accumulation of indium in the reticuloendothelial system. Generally, the soluble salts of Group III metals are converted into colloidal hydroxides at body pH, and detoxication is mainly via accumulation in the reticuloendothelial system prior to fecal excretion via the bile.

At the cellular level, detoxication of metal ions is complex, and the mechanisms can at best be speculated on. Some toxic ions, such as lead or mercury, react irreversibly with metabolites such as reduced glutathione; this reaction indirectly protects the more vulnerable and more essential enzyme and respiratory systems. Some metals are detoxicated by incorporation into inactive storage proteins. In hepatic cells, lead ions from intranuclear inclusion bodies; these bundles of lead protein complexes exert no further toxic action. Metallothionein is an inactive protein found in the liver and kidneys. It contains zinc, copper, lead, cadmium, mercury, and other metals; its synthesis is induced or enhanced in the liver when the animal is exposed to toxic metals such as lead, cadmium, mercury, and other metals.

Interactions between metal compounds potentiate or decrease the toxicity of metals; the reduction in the toxicity of vanadium by chromium, cadmium by zinc, selenium by thallium, and mercury, arsenic, silver, cadmium, and copper by selenium (Frost, 1972) are some examples. The mechanisms involved in each of these sets of metal ions are not clear. Since each pair involves one essential metal, a stoichiometric antagonism may be envisioned.

Detoxication mechanisms are effective in reducing the toxic effects of metal compounds when subtoxic amounts of the metals are previously administered to animals. More than one mechanism may be operative for the detoxication of a single metal. Metal detoxication reactions are complex and have yet to be investigated in detail.

EXCRETION

Excretion is the final removal of toxic minerals and the products of their metabolic degradation. Urine and feces (including hepatic excretion via bile) are the main avenues of excretion; exhaled air, sloughed epidermal and mucosal cells, sweat, tears, respiratory mucus, saliva, and milk serve as minor but significant avenues of excretion. Hair, horns, and hooves or nails can be considered indirect avenues of excretion, because some toxic substances are immobilized permanently in these tissues and removed from metabolizing tissues of animals.

Metals are found in urine in the form of either simple ions, water-soluble salts, complexes with amino acids, or chelates. When synthetic chelating agents, such as EDTA, are administered to remove toxic metals from the body, soluble metal EDTA chelates are excreted in the urine. Sometimes the metals may be excreted as the salts of uric acid, hippuric acid, and creatinine.

Feces contain insoluble metal compounds that are not absorbed from the digestive tract and that do not undergo any metabolic change or degradation; these compounds include insoluble metal oxides, sulfates, phosphates, and others. Some soluble metal compounds, such as gallium nitrate, are converted into insoluble polymeric forms of hydrated oxides in the gastrointestinal tract following hydrolysis, olation, or hydroxylation. Some metal toxicants that enter the bloodstream are converted into insoluble or colloidal forms in the blood and taken into the reticuloendothelial system, especially the liver; most of these colloidal forms are transported through the bile into the digestive tract for final excretion in the feces. Some metal toxicants can be detoxified into less toxic gaseous products and exhaled, e.g., methyl telluride. Liver injury or malfunction intensifies such toxicities. Essential metal ions, such as sodium and magnesium, are found in significant amounts in sweat and tears. Lead and vanadium salts are found in milk. Chromium, arsenic, and other metals are found in hair and nails. The chromium and arsenic content of hair is used to assess the extent of exposure of an organism to these metals.

The inherent physicochemical properties of the metal and the physiologic response of the organism toward the metal concentration, or dose of the metal compound, both influence the route of excretion. Most studies on the metabolism of toxic metals in laboratory animals have used radioactive tracer techniques; the radioactive forms of the metal were administered in such low dosages that they always existed in ionic forms. Irrespective of the route of administration, the radioactive tracer could be found in both urine and feces; the metal is identified on the basis of its radioactivity, not by its chemical reaction. The isotopes used in toxicity studies were administered in radiocolloidal, colloidal, or particulate forms. These results lead to inconclusive evaluation of the excretion pathway for that metal and to inconsistent and contradictory reports. Chromium serves as a typical example. Radioactive chromate given in low chemical doses is found in both urine and feces; when administered with high doses of stable chromate, it appeared exclusively in the feces (Mertz, 1969).

Thus, reports about mammalian excretion of metals and their salts should be assessed on the basis of dose levels and route of administration. Certain generalizations, however, can be established. The cations of the alkali metal subgroup and the soluble oxygenated anions of the amphoteric metals of Groups IV–VII are invariably excreted in the urine. The oxidation states of the polyvalent metal ions of Groups V–VII enable them to exist as complex, hydrated, and oxygenated forms. Most potassium or sodium salts of anions are water-soluble and are readily excreted through the renal route. The water-soluble and stable chelate complexes of metals of Groups II–VIII are excreted in the urine if the organic chelate part of the molecule is not susceptible to metabolic degradation. EDTA complexes of some of the metals of Groups III–VIII are water-soluble and so stable that they are excreted in the urine following oral and parenteral administrations. The cationic salts of metals of Groups II–VIII, with the exception of beryllium, cadmium, and thallium, are excreted mostly in the feces.

The major excretion pattern of metal ions by mammals following oral uptake is summarized in Table 3-1. Small amounts of nephrotoxic cations, such as cadmium and mercury, are excreted in both urine and feces. Osmium is retained in the digestive tract; there are no reports about its excretion. Traces of cationic copper and lead are found in urine; lead from ingested tetraethyl lead is excreted in the urine. Fecal excretion of most of the ingested metals in cationic form is due primarily to poor or negligible absorption from the gastrointestinal tract, and secondarily to the ability of the liver to degrade or metabolize the nonmetallic part of complex metal salts. It is not practical to summarize here the excretion of parenterally administered metal ions.

TABLE 3-1. Major Route of Excretion of Metal Ions after Oral Ingestion by Mammals

Group		Urine	Feces
I	A	Li, Na, K, Rb, Cs	—
	B	—	Cu, Ag, Au
II	A	Be, Mg, Ca	Ba, Sr, Mg, Ca
	B	Cd, Hg	Zn, Hg
III	A	Tl	Al, Ga, In
	B	—	Sc, Y, lanthanides, actinides
IV	A	Ge as germanate	Sn, Pb
	B	Titanate	Ti, Zr
V	A	As as arsenate, Sb as antimonate, Nb as niobate	As^{3+}, Sb^{3+}, Bi^{3+}
	B	V as vanadate	V, Ta
VI	A	Se as trimethyl selenonium ion, Te as tellurate	Se Te
	B	Mo as molybdate, W as tungstate, Cr as chromate	Cr
VII	B	Mn as manganate	Mg, Re, Tc
VIII			Fe, Co, Ni, Ru, Rh, Pd, Ir, Pt

HOMEOSTASIS

Homeostasis is the ability of a living organism to maintain a state of uniformity in its internal environment and thereby in its normal body state (Cannon, 1929); this stability of fluid matrix is accomplished by a set of complex but coordinated physiologic reactions. The scope of this early definition has widened to include all types of biologic reactions that tend to maintain at a satisfactory cellular level those essential nutrients and metabolites that would be harmful at higher levels. These nutrients include essential metals; the concept and reactions are equally applicable to other metals that are chemically similar to essential metals and that are toxic to the organism. The ability of man and animals to adapt successfully to the external environment, which may contain metals in disproportionate and harmful amounts, depends on the coordinated activities of the integrated nervous and endocrine systems (Selye and Rosch, 1954). The interplay between the two systems of cellular reactions is largely responsible for the regulation and control of metals in the internal fluid matrix. Ansari et al. (1975) note species differences in zinc homeostasis.

Homeostatic mechanisms involve mainly physiologic regulation of the absorption of these minerals from the digestive tract or control of the excretion of the minerals, or both. In unusual circumstances, homeostatic mechanisms aid in sequestering or storing excess minerals in harmless conditions; some involve the quick mobilization and transport of the minerals to kidney or liver for efficient excretion; others aid in sequestering toxic metals in bone or soft-tissue organelles. Homeostasis of a metal ion comprises all aspects of its metabolism in the living animal body. This includes absorption, transport, distribution to various tissues that may require the metal for their normal activity and that may be the prime targets for the toxic action of the metal, and, finally, the different phases of the excretion of the metal ion. The physiologic pathways and what is known of the biochemical metabolism of each metal in man and animals are reviewed in Volume 2; homeostasis of individual metals will therefore not be discussed here. The principal concepts of absorption were reviewed in Chapter 2, and absorption of individual metals will be reviewed later; however, homeostatic systems involved in absorption and excretion of a few metals with established homeostasis will be briefly reviewed.
    Iron is a typical example for the involvement of the intestinal mucosa in homeostasis through control of absorption. A portion of dietary iron is solubilized and reduced to the ferrous state by the gastric juice. Compounds such as ascorbic acid, hexoses, complex organic acids, and sulfur-containing amino acids form chelates with ferrous ion. The iron-stabilization factor present in normal gastric secretion is evidently an endogenous chelate and is yet to be identified chemically. This factor helps in the solubilization of iron and partially prevents the precipitation of iron at the more alkaline pH of the small intestine. Iron enters the brush border by passive diffusion and is transferred from the serosal surface to plasma transferrin by an energy-mediated process. The iron that cannot be transferred to plasma is bound to apoferritin, remains in the mucosal cells, and is desquamated after 2–3 days. Ferritin appears to be an important mediator in iron absorption. Although it is yet to be completely identified and unequivocally established, the mechanism for the regulation of iron absorption is postulated. The columnar cells formed in crypts at the base of villi contain variable amounts of transferrin-derived iron or iron in a form capable of combining with transferrin. This intracellular deposit of iron regulates, within homeostatic limits, the amount of intraluminal iron that can enter the mucosal cells. The iron absorbed from the brush borders into the mucosal cells can either be transferred to blood plasma according to the needs of the body, or remain within the mucosal cells, where it inhibits further absorption from the intestine. This cellular iron is lost when the cells are sloughed from the tips of the villi at the end of their 2–3-day life

span. In iron overload, the mucosal cells are formed with enough iron from transferrin to limit absorption from the duodenum; this iron is lost when desquamation occurs, and is excreted. In iron deficiency, little iron will be incorporated in the mucosal cells from transferrin, and this cellular iron deficiency enhances the absorption of iron from the intestine (Forth and Rummel, 1973).

Renal excretory mechanisms involve glomerular filtration, tubular reabsorption, and tubular excretion. Homeostasis of magnesium is operative partly through renal excretory mechanisms. In normal human subjects, 3–5% of the filtered magnesium is excreted. In magnesium deficiency, reabsorption by the tubules may become almost complete, resulting in very little magnesium excretion in the urine. In magnesium overload, the excretion of magnesium increases sharply, and is attributed generally to the action of increased levels of hormones secreted from the parathyroid, adrenal, and pituitary glands. These hormones also help to mobilize magnesium from muscles and bones during magnesium deficiency. Magnesium homeostasis is diagrammed in Fig. 3-1 (Aikawa, 1971).

Hormones and metabolites help in the mobilization, transport, and distribution of essential metals, maintaining the required concentration of these nutrients for the normal functioning of tissues and organs.

Other tissues involved in homeostasis of metals through excretion include the lungs, alimentary tract, mammary glands, lacrimal glands, and skin. Epidermal desquamation provides another satisfactory excretory

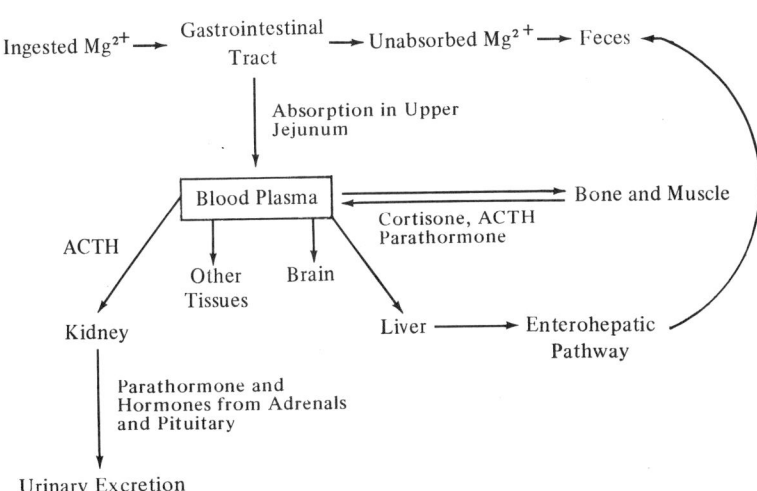

FIG. 3-1. Magnesium homeostasis.

mechanism for cationic salts, e.g., sodium, potassium, magnesium, calcium, and iron. Growing hair and nails serve as minor vehicles of homeostatic excretory mechanisms by sequestering excess metal nutrients, such as chromium, or toxicants, such as mercury, in keratinized products.

Most metals appear to exhibit limited homeostasis. With few exceptions, however, information about specific mechanisms of homeostasis for most metals is only suggestive or speculative on the basis of efficient excretion, poor accumulation in the body with age, or constancy of total-body levels, or any combination of these.

The concepts above outline at least the tip of the iceberg comprised of the intricacies of the homeostatic control of metal concentration in cells and their internal milieu through absorption, detoxication, and excretion.

4

TOXICOLOGIC SIGNIFICANCE OF THE PHYSICOCHEMICAL PROPERTIES OF METALS

The toxicity of a metal or its compounds in a biologic system is influenced by a number of factors: (1) the dose of the toxic metal, (2) the intrinsic toxicity of the metal, (3) the combining capacity of the metal, (4) the action of the biologic system to absorb and transport the metal to the target organ most susceptible to the metal intoxication, (5) the capacity of the metal to undergo biotransformation to a less toxic or a more toxic form at the target organ or during transfer, (6) the ability of the metal to bind to essential macromolecules, and (7) the homeostatic mechanisms of the organism to either excrete or sequester the metal or both. Large doses of some nontoxic metal compounds may interfere with normal cellular or physiologic processes by nonspecific activity such as changing the osmotic pressure and pH or physically changing the gastrointestinal tract.

Some of the more sensitive and susceptible biologic activities and systems affected by metal intoxication are: (1) permeability of the membranes of the cell and the subcellular organelles, (2) structure and function of nucleic acids and proteins, (3) release of potent substances such as histamine, and (4) biosynthetic formation of hormones. The defensive homeostatic mechanism of the cells and tissues combats metal intoxication either by sequestering the metal in a harmless way or by enhanced and rapid excretion of the toxic metal. The toxic metal can slow essential processes to eventual collapse and death of the organism; it can also activate and enhance certain metabolic functions, leading to uncontrolled cellular growth or tumors. Some highly toxic metals can also be stimulatory to some essential life processes at very low concentrations.

The interaction between toxic metals and susceptible systems in biologic tissues is usually nonspecific and complex. The intrinsic or inherent toxicity of a metal is a measure of its ability to interfere with the dynamic equilibria of living systems by binding strongly or irreversibly with tissues, cells, organelles, or biologic macromolecules involved in essential and dynamic processes. A study of the physicochemical properties of a metal and its interaction with biologic macromolecules will explain its inherent toxicity and its overall toxicity in biologic systems. Metal toxicity is associated with the physical and chemical properties of the metals, which differ according to their position in the periodic table (Pierre-Bienvenu et al., 1963). It is difficult to associate the toxicity of a given metal with its group in the periodic table, but the toxicity can be related to its position in the horizontal period of the table. The physicochemical properties involved in metal toxicity are: (1) the electrochemical character and the oxidation state of the metal; (2) the particle size of the metal or compound, especially in inhalation toxicity; (3) the solubility and stability of the metal compounds in biologic fluids and tissues and the degree of hydration of the ions formed; (4) the extent of hydrolysis of metal salts and subsequent olation in the tissues and the solubility and reactivity of these products; (5) the tendency of the metal compounds to exist in radiocolloidal (also designated microcolloidal), colloidal, and particulate forms in the tissues; and (6) the susceptibility of these metal compounds to be sequestered, metabolized, detoxicated, and excreted or harmlessly sequestered within the cell through chelation to cellular constituents.

The electrochemical character of a metal is directly involved in the capacity of the metal and its ions and compounds to coordinate or chelate with electron-donating oxygen, nitrogen, and sulfur atoms of biological ligands or biologic macromolecules, cellular membranes, subcellular particles, and tissue components, and in the stability of these metal chelates within living tissues. The electron-donating oxygen, nitrogen, and sulfur atoms of macromolecules play a key role in the formation and stability of metal coordination or metal chelate compounds. The stability of these bondings influences the toxicity of the metal involved in the chemical bond. The oxygen and electron transport systems involve essential metals such as iron, copper, and vanadium. These systems bind oxygen securely under certain environmental conditions, but release the oxygen under different conditions and maintain the dynamic continuity of function, flexibility, and turnover. Other metals bind oxygen in an irreversible and stable condition. Toxic metals fix biologic systems and structures into irreversible and inflexible conformations. It should be remembered that only reversible biochemical reactions can maintain the orderly, dynamic condition of living cells. Nature and nurture contribute to the homeostatic capacity of the organism in dealing with the toxic metals and therefore indirectly influence the toxicity of metals.

PHYSICOCHEMICAL PROPERTIES OF BIOLOGIC SIGNIFICANCE

Electrochemical Character and Electronic Configuration

Within certain groups of similar metals, toxicity is associated with the electrochemical character of the metals, especially the electrode potential, ionization potential, and electropositivity or electronegativity. Electropositivity can be defined as the relative ability to lose electrons to attain stable electronic configuration. Bienvenu et al. (1963) observed a relationship between toxicity of metals and periodicity of elements in the periodic table. Toxicity decreases with stability of the electronic configuration. Metals of subgroups IA and IIA are strongly electropositive; these metals form mostly electrovalent compounds and exist as free cations in biologic tissues. Within subgroups IA and IIA, electropositivity increases with a concomitant decrease in electronegativity, and ionic radius increases with an increase in the atomic number. The toxicity of these metals increases directly with electropositivity and atomic number: Na < K < Rb < Cs; and Mg < Ca < Sr < Ba. The lighter metals lithium and beryllium are less electropositive but more toxic than the other members due to their small ionic radius and higher charge-to-mass ratio, factors that are responsible for penetration and diffusion into tissues. Within certain subgroups, i.e., IB, IIB and IIIA, the toxicity of the metal also increases with electropositivity: Cu < Ag < Au; Zn < Cd < Hg; and Al < Ga < In < Tl. This increase can be attributed to the affinity of these metals for amino, imino, and sulfhydryl groups, which are active sites on a number of enzymes (Danieli and Davis, 1951; Somers, 1960). This generalization cannot be extended beyond Group IV; the electropositive character gradually decreases with a concomitant increase in electronegative character after this group. The metals of Group IV and beyond form mostly strong covalent compounds and coordinate and chelate complexes with biologic ligands. Some form oxyacids with the metal as a part of the anion. The stability of these types of covalent linkages increases in the following order: IB < IIB < IIIB < IIIA < IV < V < VIII. The gradual increase in the intraperitoneal LD_{50} values of water-soluble metal salts belonging to these groups substantiates this observation. With metals of Group VI, VII, and, to some extent, Group VIII, the electropositivity gradually decreases with a concomitant increase in the intraperitoneal LD_{50} value, especially for the cationic salts. This generalization is not applicable to anionic salts in which the metal is a part of the oxyacid, e.g., MoO_4, MnO_4^+, CrO_4^{2-}.

Toxicity can be related to stability of electronic configuration for some metals; however, this cannot be extended to all of them. The alkaline earth metals, transition metals, and metals that form stable compounds resulting in a full complement of electrons in s,p,d-orbitals of outer energy shells can be considered as relatively low toxic metals. In Group IV, the toxicity of elements increases: C < Si < Ge < Sn < Pb. The electronic configurations of the compounds of these elements are the stable s,p; s,p^3; and s,p^2. Substantial energy is required to break these configurations; a diamondlike structure is present with silicon and germanium compounds; with tin, the chances for a stable compound or complex with s,p^3 configuration is decreased. An increased number of nonlocalized electrons in lead gives rise to ready solubility, high reactivity, and electrical conductance, and also to increased toxicity. With Group V metals, the s^2,p^3 electrons can attain stable configuration in two ways: (1) stable doublet with s^2,p^6 configuration corresponding to the inert gases, and (2) s^2,p^3 can change to s^2,p^2 and to the more stable s,p^3 of transition metals. The abrupt increase in arsenic toxicity, antagonistic to phosphorus, may be interpreted as being due to the weakening of the stable s,p^3 configuration in a biologic medium. With antimony, the electrons liberated during the transition $s^2,p^3 \rightarrow s,p^3$ can be localized in the empty $4f$-orbitals; hence, acute antimony toxicity is less well displayed than is that of arsenic; however, the chronic toxicity of antimony still persists. Bismuth salts are not as highly toxic as arsenic or antimony because the transition or accommodation of liberated electrons in $5f$-orbitals is easier and does not require large amounts of energy, and the stability of d- and f-orbitals increases with the principal quantum number.

In biologic fluids, the biologic availability and activity of a metal species depends on its oxidation state and the rapidity with which the metal ion can undergo oxidation and reduction. Divalent iron is more easily utilized than Fe^{3+}; in fact, Fe^{3+} is reduced to Fe^{2+} during absorption in the gastrointestinal tract. The toxicity of some heavy metals can be associated with their state of oxidation. The higher oxides of manganese, vanadium, molybdenum, lead, and barium, and the polyvalent oxyacid salts, such as MoO_4^{2-}, MnO_4^-, and VO_4^{2-}, are more toxic than the corresponding lower oxides or lower oxyacid salts (Levina, 1966). These higher oxides may undergo spontaneous reduction to the lower forms, disrupting the delicate mechanism of cellular electron transport systems. The lower oxides of metals such as arsenic, antimony, and cobalt are more toxic than their stable higher oxides; this could be due to the tendency of these lower oxides to become oxidized to the stable, higher-valence states, disrupting cellular processes.

Particle Size

Coarse dusts of oxides of aluminum, manganese, and their salts are harmless following inhalation, but particles of these oxides less than 0.2 μm in diameter produce pneumonitis. Asbestos, a polymer containing silica and magnesium oxide in large amounts and traces of iron, chromium, cadmium, and zinc, is harmless, but inhalation of fine fibers of asbestos, 0.5 μm in diameter and about 80 μm in length, causes cancer irrespective of the chemical composition of the asbestos. The small size of these particles allows greater deposition of these toxic particles in the lower region of the lungs and eventual absorption into the bloodstream, as detailed in Chapter 2.

Solubility of Metal Compounds and Hydration of Metal Ions

The solubility of metals and their compounds in water and in lipids considerably influences their biologic availability, utilization, and toxicity. Metallic mercury is absorbed through the skin due to its solubility in lipids. Simple salts of essential trace metals and macrometals are generally soluble in water. The harmful effects of a toxic metal are more pronounced with its soluble salts than with its sparingly soluble salts; insoluble oxides are less toxic than the more soluble chlorides or nitrates of the same metal. The toxicity of the salts of a toxic metal can be expressed as follows: nitrates > chlorides > bromides > acetates > iodides > perchlorates > sulfates > phosphates > carbonates > fluorides > hydroxides > oxides. The solubility of metal salts in water is generally high with the metals of Group I, and decreases progressively to Group IV, where metal salts show minimal aqueous solubility. The anionic salts of metals in subsequent groups show progressive increase in aqueous solubility, reaching a maximum for salts of metals of Group VII. Within each group, the aqueous solubility of metal salts decreases with increase in the atomic number. There are solubility differences between each pair of subgroups. The solubility of metal salts decreases progressively with successive horizontal periods in the periodic table. Salts of metals belonging to the first three periods are more soluble in water than the metals of the subsequent periods. Metals of the sixth period are potentially the most toxic of the elements in the periodic table, but the poor solubility of their salts masks their inherent high toxicity. The toxicity of such metals increases in direct proportion to their aqueous solubility. The fairly high solubility of some salts of lead, mercury, and thallium renders these metals and their compounds highly toxic.

Passive diffusion, facilitated diffusion, and active transport are the mechanisms involved in the transfer or absorption of metal ions across cell membranes or the intestinal wall. Hydration increases the size of ions and influences their absorption.

The ion–dipole interaction enables the metal ion to acquire two sheaths of water of hydration, a firmly attached primary and a less firmly attached secondary layer. The degree of hydration is proportional to the size and charge of the metal ion; the smaller the size and the higher the charge, the greater will be the hydration. Lithium has a great amount of water of hydration because the shorter cation dipole distance more than compensates for the stronger attraction between the lithium ion and an anion. Liquid water is considered to be an equilibrium mixture of monomeric water and "flickering clusters" of many molecules joined together in an icelike structure (Nemethy and Scheraga, 1962a,b). Liquid water is present in the lipoprotein pores and the charged lining of the pores of the intestinal wall. Inorganic ions with intense electric fields at their surface, when put in water, loosen and break up the adjacent water structure by ion–dipole interaction. Charged metal ions can break up the structure of water in the pores of the biologic membrane and facilitate ion absorption or transfer. Both the size of the hydrated ion and its ability to break water structure contribute to the transfer of charged metal ions across biologic membranes.

Hydrolysis and Olation

Hydrolysis of soluble metal salts at body pH can either potentiate or decrease the inherent toxicity of a metal. Although the pH of biologic fluids in mammals is usually slightly alkaline (pH 7.4), the pH of the fluids in the mammalian gastrointestinal tract varies from pH 2 to pH 6 in the stomach, and averages pH 6.8 in the intestines—it goes above pH 7 under certain conditions. At the pH of the living systems, soluble metal salts of Groups I and II undergo rapid ionization and exist as ions. Soluble metal salts of Groups III–V undergo rapid ionization and hydrolysis to insoluble metal hydroxides. The precipitation of these hydroxides depends on their solubility, coprecipitation by other ions, and the pH of the biologic fluids. The initial reaction is represented:

$$Ce(NO_3)_3 + 3H_2O \rightleftharpoons Ce(OH)_3 \downarrow + 3HNO_3$$

If the hydrolysis occurs in the intestines, the insoluble hydroxides are sparingly absorbed; a metal that is toxic in the ionic form is thus converted into a relatively less toxic particulate form. If the hydrolysis occurs in the blood and other tissue fluids, the situation is different. In the ionic form, the metal may be bound to some serum protein fraction and cleared rapidly from the blood, rendering the metal less toxic; the metal hydroxide in the particulate form would remain in the blood for a longer time before removal by phagocytosis to the reticuloendothelial system and would be more harmful. Subsequent reactions of these forms of metal species within the tissues are slow and depend on the basicity of the metals, their charge, and their oxidation state. These reactions include chelation with organic ligands and olation.

Either or both olation and oxolation of the products of hydrolysis in biologic fluids can lead to the polymerization of the hydrolytic products, which are more sparingly soluble forms; thus, toxic metal compounds are transformed into comparatively nontoxic or less toxic forms. Soluble salts of metals such as zirconium, hafnium, and niobium are rendered less toxic by this olation and oxolation. Soluble bismuth salts such as bismuth nitrate are converted into insoluble and less toxic oxysalts such as bismuthoxynitrate. Olation and oxolation are both illustrated by the following specific example and general equations:

$$Ga(NO_3)_3 \xrightarrow{H_2O} \underset{\text{hydrated compound}}{Ga(NO_3)_3 \cdot 3H_2O} \xrightarrow{H_2O} \underset{\text{hydroxylated product}}{[Ga(OH)(NO_3)_2 \cdot H_2O]_n} + HNO_3$$

$$\downarrow$$

$$\underset{\text{oxolated aggregate}}{[GaONO_3]_n}$$

The general mechanism is envisioned as follows:

$$M(NO_3)_3 \rightleftharpoons M^{3+} + 3\,NO_3^-$$

$$\downarrow +H_2O$$

$$\left[\begin{array}{c}\\ =M\\ \end{array}\underset{H_2O}{\overset{H_2O}{<}}\right]^n \rightleftharpoons \left[\begin{array}{c}\\ =M\\ \end{array}\underset{H_2O}{\overset{OH}{<}}\right]^{n-1} + H^+$$

$$\searrow$$

$$\left[=M\underset{OH}{\overset{OH}{<}}M=\right]^{2n-2} + H^+$$

$$\swarrow$$

$$\left[=M\underset{O}{\overset{O}{<}}M=\right]^{2n-4} + 2\,H^+$$

The acid formed is often toxic.

Colloidal and Radiocolloidal Behavior

Some metal salts that are soluble at acidic pH precipitate with the change in pH of the medium. Especially at the near-neutral pH of biologic fluids, some salts of metals of Groups III–VII are hydrolyzed to their insoluble hydroxides. When the concentrations of metal salts and their hydroxides exceed their aqueous solubility, the compounds separate as precipitates; as insoluble particulate material, these compounds will be poorly absorbed from the gastrointestinal tract. At concentrations just below the limits of aqueous solubility, these metal compounds segregate as either colloidal or radiocolloidal (microcolloidal) forms, instead of remaining in true solution. This phenomenon occurs especially when they do not undergo further olation or oxolation. The radiocolloid, or microcolloid, consists of the insoluble metal hydroxides existing as minute colloids at concentrations *far below the solubility constant for the ion*. In biologic fluids and tissues, the microcolloids are either minute particulate materials with these metal ions absorbed on them or insoluble metal hydroxides, or both (Schweitzer and Scot, 1955). This phenomenon takes place at such exceedingly low concentrations that their presence in biologic fluids could be detected only by radioactive tracers. Hence, these forms were named *radiocolloids*—a misnomer, since they are not necessarily radioactive; they are often called *microcolloids*. These colloidal, or radiocolloidal (microcolloidal), forms are stabilized by aggregating around a minute particle or by a protective coating of some protein constituent of body fluids, or by both means. The limited absorption of these forms from the gastrointestinal tract differentiates them from precipitated forms, and their subsequent behavior depends on the aggregate size. The absorption of microcolloids and colloids is much less than would be anticipated were true solutions involved. This phenomenon occurs with the heavier metals of Groups III–VIII in varying degrees and has been studied in detail with the lanthanides (Schweitzer, 1956).

When the lanthanide cations are injected into blood plasma, they separate as insoluble hydroxide particulates if the concentration is above the molar solubility of these hydroxides, i.e., 8.8×10^{-5}M. The lanthanide hydroxides separate as colloidal forms if the concentration ranges from 8.8×10^{-6} to 5×10^{-7} M; aggregates of metal hydroxides become surrounded by a protein coat or similar peptizing agent. If the concentration of the injected lanthanide salt is far below the molar solubility of the lanthanide hydroxide—: i.e., if it ranges from 10^{-7} M to 10^{-11} M—the lanthanide hydroxides separate in radiocolloidal forms; these forms exist as a filterable and centrifugable disperse phase. This formation is due to the polymerization or coalescence of these hydroxides at the pH of the blood. The polymerization increases with decreased basicity of the hydroxides. The plasma proteins interact with these aggregates to stabilize them, or these aggregates adsorb to any particulates. It should be remembered that this colloidal character confers great adsorptive capacity for sparingly soluble compounds. Diffusion coefficients of radiocolloids are much less than those of the metal compounds in their ionic form; thus, radiocolloids will not pass through membranes permeable to ions.

While the removal of precipitated metal compounds from the blood by phagocytosis is rapid, the removal of colloidal forms is slower, and that of the radiocolloidal forms is the slowest; the removal of the ionic forms is, of course, the most rapid. The distribution of these forms between blood and other tissues differs; while the ionic and radiocolloidal forms of the lanthanide hydroxides are concentrated more in the bones, the colloidal and precipitable forms are deposited more in the reticuloendothelial system. The toxic symptoms depend on this distribution; deposition in the skeleton does little harm, while deposition in the liver causes hepatic damage.

Metals of Groups III—including the lanthanides and actinides—IVB, and VB exhibit this radiocolloidal behavior, while metals of Groups I, II, IVA, VA, VI, VIIA, and VIII show little or none.

HARD AND SOFT ACIDS AND BASES THEORY

Metals and their ions have the capacity to combine with functional groups or ligands present in biologically active molecules such as $-SH$, $-OH$, $-NH_2$, $-COOH$, $-PO_4^{2-}$, $-O^-$, and $-O-O$. The toxicity of a metal depends on the stability of this metal–ligand bond and the kinetics of formation and degradation. The ligands also contain dissociable protons that compete with the metals for the ligands as do protons from water. The relative affinity of metals and their ions toward other anions and carboxyl groups can be assessed; the affinity of metals to thiol groups is difficult to assess. The binding of metals with organic $-OH$, $-PO_4H_2$, and $-COOH$ groups is of the strong electrovalent type. The $-OH$ group of water complexes with metals or their ions, and the electrochemical behavior of metal $-OH$ complexes differs from that of simple hydrated metal cations. Metals form coordinate covalent complexes with nitrogen-donor ligands or with amino groups. The relative affinity of some metals toward the amino group has been assessed, and the metals can be arranged on the basis of this affinity in this decreasing order: $Hg > Cu > Ni > Pb > Zn > Co > Cd > Mn > Mg > Ca$. The specific affinity is the individual characteristic of the metal and depends also on the macromolecule that carries the $-NH_2$ group.

The affinity of metals toward the –COOH group can be arranged in this decreasing order: Ca > Mg > Mn > Ni. The affinity of the metal toward active ligands cannot be categorized on the basis of the periodic chart groups.

The affinity of the metals and their ions toward active groups can be predicted to some extent on the basis of the principle of hard and soft acids and bases (HSAB theory) developed by Pearson (1963, 1967, 1968); this theory explains empirically the basicity of a metal, its affinity for a particular ligand, and the bond strength between the metal and the ligand attached to a biologically active molecule. The assumptions of HSAB theory are simple: (1) If there is a bond between the two different atoms or groups of atoms, one plays the role of an acid and the other that of a base; (2) electrons hold the bonded atoms together; and (3) electron mobility decides the softness or hardness of the acids and bases.

The acid is defined as the species that accepts or has vacant accommodations for electrons, and the base as the electron-donor. On this basis, all the positively charged metal ions, atoms of nonmetals such as O_2, and the halogens are termed *acids;* nonionized water, ammonia, and negatively charged anions of nonmetals and oxyacids are called *bases*. When a metal salt is dissolved in water, the cationic portion, the acid (A), is bonded to the basic end of the water molecule; the anionic portion, the base (B), is bonded to the acidic end of water:

$$AB \rightleftharpoons A^+ + B^-$$
$$AB \underset{}{\overset{HOH}{\rightleftharpoons}} \underbrace{A(OH_2)_n}_{\text{acid}} + \underbrace{(HOH)_n B}_{\text{base}}$$

In biologic fluids and tissues, the anionic portion could be the ligand of an active macromolecule.

Electron mobility or polarizability characterizes the hardness or softness of both the acids and the bases. The species are called *soft* if electrons can be easily moved or transferred from them, *hard* if the electrons are held firmly. The characteristics of a hard acid or electron-acceptor are: (1) low polarizability, (2) high electropositivity, (3) large positive charge or oxidation state, (4) small ionic size, and (5) ability to form an ionic bond; the converse is true for a soft acid (Table 4-1). A hard base or electron-donor is characterized by the following properties: (1) low polarizability or ionization, (2) high electronegativity, (3) large negative charge or small ionic size, (4) capacity to form ionic bond, and (5) availability of high energy and inaccessible electronic orbitals. Strong and stable bonds are formed between hard acids and hard bases and between soft acids and soft bases, whereas bonds between hard bases and soft acids and between soft bases and strong acids are weak and partially unstable.

Pearson (1968) classified metal and nonmetal ions into hard and soft

TABLE 4-1. Properties of Hard and Soft Acids and Bases

Property	Hard	Soft
Acids (Electron-Acceptors)		
Polarizability	Low	High
Electropositivity	High	Low
Positive charge of the oxidate state	Large	Small
Size	Small	Large
Type of bond	Ionic	Covalent
Outer electrons on donor atoms	Few and easily excited	Several and easily excited
Bases (Electron-Donors)		
Polarizability	Low	High
Electronegativity	High	Low
Negative charge	Large	Small
Size	Small	Large
Type of bond	Ionic	Covalent
Available empty orbitals of donor atoms	High energy and inaccessible	Low lying and accessible

acids and bases on the basis of the formation constants of some of their more common simple salts. Some metal ions at their lower valence levels or oxidation states are classified at the borderline between hard and soft acids (Table 4-2). Hard acids combine more readily with ligand atoms of the second short period of the periodic table (N, O, F) than with atoms of subsequent periods:

$$N > P > As > Sb$$
$$O > S > Se > Te$$
$$F > Cl > Br > I$$

Metal ions such as NA^+, K^+, Ca^{2+}, Mg^{2+}, Al^{3+}, and Ga^{3+} combine strongly with hard bases such as water; all these ions are strongly hydrated in solutions due to this preference. Oxygen donor atoms present in groups such as –OH, carbonates, phosphates, glutamates, oxalates, and lactates, and nitrogen donor atoms from groups such as amines readily combine with the hard acids listed above. The poor absorption of lanthanides and other Group III metal ions from the digestive tract could be explained by the formation of stable and insoluble phosphate salts. Soft acids combine more strongly with atoms of the fourth period, rather than with the atoms of the second short period:

$$Sb > As > P > N$$
$$Te > Se > S > O$$
$$I > Br > Cl > F$$

Soft acids, such as Cd^{2+} and Pb^{2+}, form strongly bonded linkages with soft bases such as –SH or –SR; this ability explains the irreversible inactivation of –SH group-carrying enzymes. Hard acids such as Mn^{2+} do not inactivate thiol enzymes, since they can form only weakly bonded linkages with –SH groups. The reaction of soft acids with R–SeH, a soft base, to irreversibly inactivate selenium-dependent enzymes provides further evidence of the usefulness of the concept. Most of the essential trace metal ions are either soft acids or borderline soft acids. This property enables them to form reversible and less stable linkages with biologically active ligands that are hard bases, and thus to maintain the dynamic conditions within the living cell. Toxic metals are often either hard or soft acids that form irreversible and stable bonds with active ligands.

TABLE 4-2 Classification of Ions and Reactive Groups[a]

Group	Hard	Intermediate	Soft
		Acids	
I	H^+, Li^+, Na^+, K^+, Rb^+, Cs^+	Cu^{2+}	Cu^+, Ag^+, Au^+
II	Be^{2+}, Mg^{2+}, Ca^{2+}, Sr^{2+}, Ba^{2+}	Zn^{2+}	Cd^{2+}, Hg^{2+}, Hg^+
III	Al^{3+}, Ga^{3+}, In^{3+}, Sc^{3+}, La^{3+}, Ce^{3+} and other lanthanides except Pm, UO_2^{2+}	Y^{3+}	Tl^+, Tl^{3+}
IV	Si, Ge^{4+}, Sn^{4+}, Ti^{4+}, Ti^{4+}, Zr^{4+}, Hf^{4+}	Pb^{2+}, Sn^{2+}	
V	As^{3+}, NbO^{2+}, VO^{2+}	Sb^{3+}, Bi^{3+}	Ta
VI	Cr^{3+}, MoO^{3+}, WO^{4+}		O, Te^{4+}, RTe^+, RSe
VII	Mn^{2+}		Cl, Br, I
VIII	Fe^{3+}, Co^{3+}	Ni^{2+}, Fe^{2+}, Co^{2+}, Ru^{2+}, Rh^{3+}, Os^{2+}, Ir^{3+}	Pt^{4+}
		Bases	
	N_2O, OH^-, F^-, Cl^-, PO_4^{3-}, SO_4^{2-}, CO_3^{2-}, ClO_4^-, HCO_3^-, NH_3, RNH_2, O^{2-}, ROH, RO^-	Br^-, NO_2^-, SO_3^{2-}	SH^-, S^{2-}, I^-, RS^-, CN^-, SCN^-, CO, R_2S, RSH, RS-SeH, $(RO)_3P^-$

[a]Adapted from Pearson (1968).

COORDINATION AND CHELATION

Coordination is a type of chemical bonding between metals and ligands involving coordinate covalency. Atoms, groups of atoms, or ligands surround the metal; the atoms directly bound are called the *donor atoms,* and they are less electronegative than the metallic species. The preferential coordination of a metal ion with a given donor atom is a function of the electronegativity of the metal ion. The predominant features in this bonding are electrostatic attraction and charge redistribution to the metal. The total number of atoms attached to the central metal atom is the *coordination number* of the metal, and the spatial distribution of the atoms is called the *configuration* of the coordination complex. All participating atoms are within bonding distance of the central metal atom; some may be farther away than others, and may contribute to the coordination of more than one metal.

Chelation is a special form of very stable coordinate complexation. A single-ligand molecule with two or more electron-donating atoms such as sulfur, oxygen, or nitrogen spaced along the ligand molecule coordinates or binds to a metal with the formation of a heterocyclic ring. The ring formation is stable and contains at least one covalent bond. The ligand could be bidentate (two donor atoms) or multidentate; e.g., EDTA has four donor atoms. The metal is held very firmly in "clawlike" (*chelate,* from Gk. *chēlē,* "claw") bonds. The multivalency of the metal ion may allow for more than one chelate ring. Metal-organic complexes present in biologic systems may be either coordinate complexes or chelate complexes; both types of complexation influence biologic activity.

Metals in the form of coordination complexes are involved in a number of essential biochemical functions in living tissues: (1) biosynthesis and degradation of macromolecules such as proteins, carbohydrates, and lipids by bond formation and cleavage involving peptidases, decarboxylases, and phosphorylases, e.g., metals such as manganese, magnesium, or zinc, or metalloenzymes coordinating to amino and peptide nitrogens and to carboxyl groups; (2) maintaining the conformation of macromolecules, e.g., zinc in insulin, manganese in RNA, and iron in porphyrin complexes; (3) oxidation–reduction reactions in cellular respiration, e.g., iron in cytochromes, catalases, peroxidases, and phenol oxidases; (4) storage and transfer of essential metals needed for other metabolic reaction, e.g., ferritin; and (5) transport, removal, or storage of toxic metals, e.g., metallothionein. The metals involved in these functions are mostly transitional metals, the outer *d*-orbitals of which have about the same energy as the *s*-

and p-orbitals of the valency shell; these metals exhibit extensive complex formation in biologic media. The electron-donating oxygen, sulfur, or nitrogen atoms of the ligands of the macromolecules are the coordinating atoms. If the metal achieves its maximum coordination number in the formation of a metal coordination complex, the resulting compound is more stable than compounds in which less than the maximum number of groups or atoms are coordinated. While lighter essential metals do not achieve the maximum coordination number, the heavier toxic metals do; this stability blocks essential and functional groups of macromolecules. The size of the ligands and the valency of the metal ions contribute to the stable geometric configuration associated with the coordination number.

The HSAB theory predicts in a qualitative way the stability of the metal–ligand bonds and indirectly determines the toxicity of a metal. The theory visualizes all types of bonding: electrovalent, covalent, and coordinate covalent. In biologic systems, the specificity of the binding of a metal ion to a ligand depends on coordination, i.e., the ability of the metal ion to accept the electrons donated by donor oxygen, sulfur, and nitrogen atoms of the ligands. According to HSAB theory, the ligands containing oxygen donor atoms, and nitrogen donors in NH_3, $-NH_2$, and $=NH$, are hard bases, while donor sulfurs in $-SH$, RSH, RS, and R_2S are soft bases.

Lighter metals of Groups IA, IIA, IIIB, and IVB of the second and third periods coordinate strongly with oxygen donor atoms. In the fourth period, Cu (IB), Zn (IIB), and Co and Ni (VIII) coordinate strongly with nitrogen donor atoms, but they retain the ability to coordinate with oxygen to a greater extent than do the lighter metals. Vanadium is a powerful oxygen coordinator. Beyond vanadium, the metals show an increasing tendency to coordinate with carbon; zinc combines with cyanide, and chromium, nickel, and iron form stable volatile carbonyls. In the fourth and fifth periods, the tendency of the heavier metals to coordinate with nitrogen and halogen donor atoms is greater. Platinum coordinates more with halogens than with oxygen or sulfur. If metals were to coordinate with sulfur or oxygen to form stable complexes, they would be more toxic. Small metal ions with high charge-to-density ratios, e.g., lithium and beryllium, form stable and irreversible coordination complexes. The capacity of the heavy metals to form stable and irreversible coordination complexes with oxygen, sulfur, and nitrogen donor atoms of the ligands of biologically active macromolecules may be responsible for their toxicity.

Metal chelation in biologic systems has wide implications and is involved in a number of essential functions, the most important of which is the regulation of metal ion concentration. Mosses and lichens growing on barren rock extract essential trace metals from the rock by generating chelating agents (Schatz, 1963). Soybeans growing in alkaline soil solubilize the iron in the soil by secreting chelating agents (Rollinson, 1969). Some *Salmonella* mutants unable to synthesize chelating agents utilize the chelating agents produced by other microorganisms to extract and transport iron into their own cells (M. Luckey *et al.*, 1972). Most metal chelates are highly soluble in tissue fluids at biologic pH, and hence are easily absorbed, transported, and excreted. In mammals, ferroxamine is a specific agent for ferric iron, capable of extracting iron from all tissue components except ferritin or transferrin; it is synthesized in the body during excessive iron ingestion or iron intoxication to aid in rapid iron excretion. Its lipid solubility enhances nonrenal excretion, and its aqueous solubility aids its movement into intra- and extravascular spaces (Waxman and Brown, 1969). The major function of metal chelation is to enhance the absorption of essential metals when the levels of the metals become low and their presence critical for essential life processes. Other nonessential and toxic metals that are similar to essential metals in electronic configuration and chemical properties undergo the same chelation reaction and gain rapid entry into living systems to exert their toxic effects.

Other important functions of metal chelates in mammalian systems include: (1) transfer of material, e.g., iron by transferrin and oxygen by hemoglobin and hemocyanin; (2) transfer of energy, e.g., electron transport by cytochrome system; (3) storage of essential metals, e.g., iron by ferritin and copper by ceruloplasmin; (4) stereochemical rearrangement of a biological macromolecule (chelate), causing either masking or unmasking of active centers, e.g., enzyme activation and inactivation; (5) macromolecule formation due to polymerization; (6) increasing the liposolubility of the chelating molecule, which could be an active metabolite, to enhance its penetration into the living cells; and (7) detoxication and excretion of toxic metals. Toxic metals, following their entry into living systems, compete with benign and essential metals and strongly inhibit these functions. Transferrin in the blood plasma is utilized by toxic metals for their transport and distribution in the tissues, and organisms are thereby deprived of essential iron. Previous reviews on biologic aspects of chelation dealt with the detoxication and removal of radioelements from living systems using synthetic chelating agents.

There is a large variety of naturally occurring chelating molecules in mammalian systems:

1. Hydroxy acids: citric acid, other hydroxy acids of the Krebs cycle
2. Hormones: thyroxine, cortisone, histamines
3. Conjugated diketones: ascorbic acid, acetoacetate

4. Vitamins: cobalamines, pteridines, riboflavin
5. Porphyrins: hemoglobin, hemocyanin, catalase, cytochrome C oxidase
6. Amino acids: histidine, serine, cysteine
7. Polyamines: spermine, spermidine, catecholamines, epinephrine, norepinephrine, DOPA analogs
8. Proteins: metal-activated enzymes, metalloenzymes
9. Nucleic acids: nucleic acids, nucleoproteins and their derivatives

Chlorophyll is the major metal chelate derived from plant sources. Metal-chelating compounds derived from microbic sources and used therapeutically in mammalian systems are: (1) tetracyclines, (2) penicillin and penicillamine, (3) mycobactin T, (4) ferrioxamine B, (5) itoic acid, and (6) pyrimine. Important synthetic chelating agents are: (1) ethylenediamine tetraacetic acid (EDTA), (2) diethylenetriamine pentaacetic acid (DTPA), (3) hydroxyethylethylenediamine tetraacetic acid (HEDTA), (4) ethylenediamine-*bis*(1-hydroxyphenyl)acetic acid (EDDHA), (5) isoiazid, (6) salicylate, and (7) EGTA.

Di-, tri- and tetravalent metal cations form complexes with several of these chelating agents. The chelating agents possess ligands containing two or more electron-donating atoms such as oxygen, nitrogen, and sulfur. If the metal chelates with two or more of the same ligands, it forms a true chelate; if two or more dissimilar ligands are involved, it is called a mixed chelate. The chemical factors involved in chelation reactions are: (1) ring formation, (2) dentation, (3) resonance, (4) pH, and (5) specificity. The binding of the metal to the electron donors of the ligand forms a heterocyclic ring with at least one covalent bond; stable chelate rings contain 5–7 atoms. If a covalent bond is involved, chelation is restricted to the divalent alkaline earth metals, transition metals, lanthanides, and actinides. If the ligand possesses two electron-donor atoms actively participating in the chelation, the chelating molecule is called bidentate. On this basis, EDTA is multidentate. The pH of the medium should be near neutral or higher for efficient chelation, since the metal ions have to compete with protons from these ligands for the electrons that are donated by the donor atoms. In most cases, oxygen, nitrogen, and sulfur atoms of the ligands are already sharing their electrons with hydrogen atoms, and the metal must compete for them. Resonance stabilizes most of the unsaturated chelate rings. It is difficult to associate the specificity of one metal to a ligand; most chelating agents show an approximately similar order of preference for the metals. The metal chelation is more pronounced with decreased ionic radius and higher valency; the univalent ions show the least avidity, the divalent ions medium avidity, and the trivalent ions the most avidity toward the chelating molecules. The order is: $K^+ < Na^+ < Li^+ < Ca^{2+} < Mg^{2+} < Mn^{2+} < Fe^{2+} < Co^{2+} < Ni^{2+} < Cu^{2+} < Fe^{3+} < Cd^{2+} < Zn^{2+} < Pb^{2+} < Al^{3+} < Hg^{2+}$.

In general, metals prefer chelations with oxygen donor atoms; however, Cu^+, Cu^{2+}, Ag^+, Au^+, Au^{3+}, Ni^{2+}, Co^{2+}, Hg^{2+}, As^{3+}, and Sb^{3+} prefer sulfur to oxygen. Other metals—Mg^{2+}, Ca^{2+}, Fe^{3+}, Tl^{3+}, Mo^{5+}, and V^{5+}—prefer to combine with oxygen rather than with nitrogen.

Most naturally occurring chelates in mammalian tissues perform specific essential functions in association with the essential metals. Potentially harmful metals with physicochemical properties similar to the essential metals compete with these essential metals for the chelates and prevent the normal functioning of those chelates by forming irreversible and stable bonds. An essential metal such as iron becomes biologically very active by incorporation into the porphyrin nucleus; a potentially harmful or toxic metal might become more toxic in an analogous way. Some metal ions interfere with the activity of enzymes that are involved in the catabolism of organic carcinogenic compounds and can indirectly act as toxic metals. Most toxic metals gain entry into living cells by chelating with naturally occurring agents. This chelation usually increases the lipid solubility of the metals; otherwise, the metal ions remain on the cell surface. After entry into the cells, the organic part of the metal chelate is metabolized by lysosomal hydrolases to release the metal ion in a very active form; it can then combine with subcellular components and other macromolecules to exert its toxic effects. When the toxic metal chelate is quite stable and not susceptible to the lysosomal hydrolases, this action sequesters the metal chelate in the residual lysosomes from which it is excreted. Thus, chelate formation can either potentiate the toxicity of a toxic metal or detoxify the metal and aid in its excretion.

Chelation is useful therapeutically for the removal of toxic metals following accidental poisoning. Synthetic organic chelates having multiple polar groups and with very low or no toxicity to mammals—such as –OH, –COOH, –SH, and –NH$_2$—are used for this purpose; the metal chelate complexes should be readily excreted by the kidneys or intestines. There should be at least one free polar group on the molecule following its chelation to ensure enough water solubility for kidney excretion. If these synthetic chelates are used indiscriminately for removal or detoxication of toxic metals, they will ultimately cause the depletion of essential metals from the body, producing essential metal deficiencies. Calcium disodium EDTA is an effective remedy for lead poisoning (Bessman *et al.*, 1954); tetrasodium EDTA is effective for treating lime burns of the cornea (Gundorova *et al.*, 1967) and for treating hypercalcemia patients by removing blood calcium (Spencer *et al.*, 1956). Sodium EDTA suppresses the toxic effect of vanadium in chicks (Hathcock *et al.*, 1964), but it is not known whether it removes vanadium from the system. High doses of EDTA prove to be teratogenic in pregnant rats and lead to congenital malformation in the young by depleting the body stores of zinc (Tuchmann-Duplessis and Mercier-Parot, 1956; Swenerton et al., 1969). Desferrioxamine is used to remove excessive iron from patients suffering from iron intoxication and from idiopathic hemochromatosis (Keberle, 1964). Sodium salicylate or aurine tricarboxylic acid is used as an antidote in beryllium poisoning. DTPA can be effectively used to remove plutonium and other transuranium metals; it is used also for removing the lanthanides. The nontoxic penicillamine is effectively used to remove excess copper in patients suffering from Wilson's disease; the stable copper penicillamine chelate complex is readily excreted in the urine. The toxic dimercaprol in nontoxic doses (2–3 mg/kg body weight) can be used to eliminate mercury, gold, arsenic, and antimony in metal poisoning. The bactericidal action of tetracyclines is attributed to their chelation properties; by chelating with magnesium, the metal–tetracycline complex accumulates in the bacterial cell and prevents the normal functioning of ribosomes by dissociating them.

Chelation is an effective reaction that governs the biologic activity of metal ions in living systems. The effectiveness of chelation and other coordination reactions provides understanding of metal–protein, metal–nucleic acid, and metal–membrane interactions.

PROTEIN–METAL INTERACTIONS

Metal ions interact with amino acids and proteins in biologic systems. Histidine and cysteine bind metals more securely than do other amino acids, histidine through its imidazole nitrogen and cysteine through its thiol group. Metals are bound to proteins by the histidine and cysteine residues in the protein. Peptides and proteins have less affinity for metals than do amino acids because the amide bond proton does not ionize readily, although the metal is usually bound between this group and a terminal $-NH_2$ group.

Metals such as magnesium and calcium with empty orbitals, and metals such as Mn^{2+} and Fe^{3+} with half-filled d-orbitals, form coordination compounds with proteins. The stability of the bonds is governed by electrostatic attractions; it is increased by greater charge and smaller radius of the metal ion. Metals such as Co^{2+}, Co^{3+}, Ni^{2+}, Cu^{2+}, and Zn^{2+} form covalent linkages with proteins, and the stability of the bonding does not depend on the increasing charge or decreasing size of the metal ions. The affinity of manganese and iron toward protein ligands is: $RCOO^- > -NH_2$

> RSH > H₂O. They exhibit greater affinity toward oxygen than toward nitrogen. The affinity of cobalt, nickel, copper, and zinc is as follows: RS^- > RNH_2 > $-OH^-$ > H_2O. Zinc binds with unidentate ligands with poor stability. Mercury binds with –SH, imidazole, –NH₂, and ϵ-amino groups at neutral pH; each molecule of serum albumin binds 5 Hg^{2+} ions at pH 5 and 45 Hg^{2+} ions at pH 7.5. Lead binds with –COOH groups and causes the precipitation of proteins due to the insolubility of Pb–protein complexes. Protein–metal interaction is governed by: (1) $-COO^-$, –SH, and imidazole groups; (2) competition between the metal ions and protons for the electrons donated by the ligand; (3) desolvation of the active site and the metal ion; and (4) the tertiary structure of the protein.

In the native condition the secondary and sometimes the tertiary and quaternary structure of proteins is governed by metal ions. The metal ion may ensure the correct tertiary folding of the protein; this folding can bring together two or three amino acid residues to form an active site. The quaternary structure of protein is maintained partially by metals, e.g., zinc in insulin. If zinc is removed from the alkaline phosphatase of *Escherichia coli,* the native conformation gives way to a random coil structure. This ability of metal ions to alter the protein structure influences essential functions such as enzyme catalysis, zymogen activation, transport processes, and gas exchange. About 12–14% of all known enzymes need either a metal cofactor or a tightly bound metal atom for their activity. The lowering of the free energy of activation is attributed to the electronic deformation in the substrate, mediated by the metal; the metal is initially activated by the protein which withdraws electrons from it. Metal activation of enzymes is often ascribed to the stabilization of secondary and tertiary structures. This stabilization is significant both for the control of the catalytic activity and for the regulation of protein degradation and turnover. Metal inactivation of proteins is frequently due to nonspecific cross-linking of essential side chains, which promotes irreversible denaturation and decreases the solubility of the proteins.

Relationships between the structural and functional roles of metal ions in subunit interactions of multichain proteins are complex. Glutamic dehydrogenase in animal tissues catalyzes the reactions that link amino acid with carbohydrate metabolism and control ammonium ion concentration. This enzyme can dissociate into four subunits, and zinc is essential for both the function and maintenance of its quaternary structure (Frieden, 1958). Toxic metals with an electronic configuration similar to zinc, such as cadmium, can disrupt the function of glutamic dehydrogenase and the integration of amino acid and carbohydrate metabolism. Metals are involved in polymerization of monomer chains of proteins. The crystalline unit of insulin is composed of three dimers and two zinc atoms. The metal links the dimer through the N-terminal amino groups of phenylalanine (Marcker and Grade, 1962). Iron is involved in the polymerization of apoferritin molecules; Fe^{2+} located on the external surface of the monomer is implicated in a reversible association of ferritin subunits (Richter and Walker, 1967). Metal ions play a role in zymogen activation; Ca^{2+} binds to the β-COOH groups in the N-terminal region in the autocatalytic activation of trypsinogen to trypsin. The role of metals as allosteric effectors has not yet been studied in detail.

Gold thiomalate or thiosulfate reacts with collagen *in vivo* and improves its stability by cross-linking; the collagen becomes less soluble and less sensitive to lysosomal proteases (Adam and Kuhn, 1968). Bismuth, copper, and mercury compounds are also incorporated into collagen *in vivo*, but the effects are harmful, since this modified collagen is more susceptible to protease action. Mercury and uranyl ions readily bind to serum albumins and stabilize the albumins. Other proteins that are stabilized by metals are conalbumin, insulin, transferrin, chymotrypsin, ceruloplasmin, and ferritin.

Zymogen activation, dissociation of multichain proteins, and protein polymerization influence a number of regulatory processes and homeostatic mechanisms. Involvement of metal ions in these actions suggests an indirect regulatory role for the metals in control and homeostatic mechanisms. The various forms of metal-induced structural effects are not generally metal ion specific. A metal ion may serve two different functions in two different proteins, or even two different roles in the same protein. In some proteins, one metal ion will serve the function role, and a different metal ion will serve the structural role. Toxic metals can interfere in these processes, since the processes are not metal ion specific; thus, protein structural relationships may be major factors in the toxicity of a metal.

Toxic metals can also induce tissue damage as a result of immunologic reaction; metals can act as haptens. The toxic effects are due to both cell-mediated and humoral immunologic mechanisms. Also implicated are nonspecific paraimmunologic mechanisms similar to the Shwartzmann reaction. In each case, the response is against the organism's own protein, which is modified antigenically by the metal. The metal is either adsorbed to the surface of the protein molecule, or it changes the overall configuration of the protein. This immunologic sensitivity is considered to be genetically controlled and appears to be a simple Mendelian dominant characteristic (Polak *et al.*, 1968). Epicutaneous contact with metals such as nickel, chromium, beryllium, and mercury develops dermatitis or skin eruptions. Hexavalent chromium is converted into Cr^{3+} and is linked to the protein side chains; the compound need not remain attached to the body's protein once it has changed the protein's antigenicity (Samitz and Katz, 1964).

NUCLEIC ACID-METAL INTERACTIONS

Essential metal ions are involved in maintaining the precise conformation and structural state of nucleic acids and ribosomes for their normal biologic activity. Ribonucleic acids are functional only in the presence of monovalent or divalent metal ions. The binding of toxic metals with nucleic acids will affect the conformation and structure and may lead to impaired function.

Divalent metal ions and Al^{3+} are found in nucleic acids from a number of phylogenetically different sources. The firm binding of metals to nucleic acids suggested either chelate complexing with nitrogenous bases or sandwich complexes of ferrocene type (Wacker and Vallee, 1959). Fuwa et al. (1960) suggested the involvement of chromium in stabilizing certain regions of critical secondary structure of RNA by imposing a tertiary structure in the form of intramolecular bonds. Some metal ions, such as Mg^{2+}, Ca^{2+}, and Mn^{2+}, can reactivate a "denatured" leucyl-tRNA, which has lost its ability to bind to leucine; 6 or 7 Mn^{2+} atoms were bound to a single leucyl-tRNA molecule (Lindahl et al., 1966). Zinc ions sometimes cleave $5'$ phosphate bonds in RNA, resulting in depolymerization; in vitro RNA hydrolysis can be catalyzed by a large variety of metal ions, such as Al^{3+}, lanthanides, Cd^{2+}, Pb^{2+}, and Bi^{3+} (Bamann et al., 1954; Trapmann and Devani, 1965; Eichorn and Butzow, 1965). In aqueous solutions, the native double-stranded RNA is unstable, but can be stabilized by the addition of divalent ions, such as Mg^{2+} and Co^{2+}, which inhibit the unwinding by preventing the mutual repulsion of the negative charges of the phosphate ions. Divalent copper ions disrupt the double strand and unwind the DNA (Dove and Davison, 1962; Fuwa et al., 1960). Zinc ions at the same concentration level can unwind or rewind DNA, depending on the temperature.

In nucleic acids, the potential ligands for metal coordination in the order of availability and stability are: (1) the phosphate of the ribose phosphate backbone, (2) the oxygen and nitrogen atoms and the π electrons from the nitrogenous bases, and (3) ribose hydroxyl groups. The nucleic acids act as negatively charged polyelectrolytes due to the phosphate groups. The high-electron-density atoms of the bases are the nitrogens: N_2, N_3, and N_7 of adenine, N_3 and N_7 of guanine, N_1 of cytosine, N_1 and O_4 of thymine or uracil, and the N from the amino groups. In the double-stranded molecule, only N_7 will be available. The $-OH$ of the ribose sugar is involved in hydrogen bonding with cations. The selection of ligand binding sites by metals and the effect of this binding on nucleic acid structure and function depend on the relative affinity of the metals toward these sites. The metal binding or complexing could be simple coordination or coordination and chelation, true chelation involving two nitrogens, or mixed chelation involving one nitrogen and one phosphate. *In vitro* studies show that the phosphate group is the strongest coordinating group for some transition and most alkaline earth metals. Copper has great affinity for the nitrogen of the bases. Uranyl ions bind mostly to ribose groups. The affinity of the phosphate binding metals toward base binding in DNA decreases as follows: Mg > Co > Ni > Mn > Zn > Cd > Cu. Silver also binds to $-NH_2$ groups. The stability constant of complexes of DNA with ions such as Cu^{2+}, Cd^{2+}, and Fe^{2+} is similar to the complexes of AMP with these metals, and is less than the stability constant of ATP–metal complexes (Bryan and Frieden, 1967). The stability of the complexes increases with increased phosphate content. *In vitro* studies reveal that the relative base affinity of Hg^{2+} is as follows: thymidine > cytidine > adenosine > guanosine. Metal ions influence several fundamental processes of nucleic acid metabolism: (1) replication, (2) transcription, (3) translation, (4) cleavage of pyrophosphates, and (5) formation of phosphodiester linkages.

DNA polymerase involved in replication requires the interaction of a series of proteins with DNA template, the four deoxynucleotide triphosphates, and magnesium ions; the substitution of Mn^{2+} for Mg^{2+} leads to incorporation of both deoxyribonucleotides and ribonucleotides into DNA, causing serious errors in DNA replication. Divalent zinc and cobalt can substitute for Mg^{2+} in *in vitro* studies; other metals cannot replace Mg^{2+}. Toxic metals such as Be^{2+} evidently interfere in DNA replication by their stabilization of ribonucleotide triphosphates. DNA polymerase contains two tightly bound Zn^{2+} ions per molecule; zinc binds the enzyme to DNA, while magnesium binds the deoxytriphosphates to the enzyme. The copying of the DNA code into *m*-RNA, which utilizes ribonucleotides, is catalyzed by RNA polymerase systems. Mammalian liver RNA polymerase systems are activated by Mg^{2+} and Mn^{2+}; the Mg^{2+}-activated fraction is inhibited by Cu^{2+}, Cd^{2+}, and Co^{2+}, and the Mn^{2+}-activated fraction is inhibited by Hg^{2+}, Cd^{2+}, and Cu^{2+}. Both replication and transcription require the unwinding of double-stranded DNA and its subsequent rewinding; essential metal ions such as Mg^{2+} and Mn^{2+} are involved in the two processes. The effect of similar metal ions such as Be^{2+} and Sr^{2+} in high concentrations on the pyrophosphate cleavage in this replication would be highly inhibitory and detrimental; the inhibition of DNA replication in mice and rabbits by beryllium has been confirmed. Beryllium is bound to an unidentified nuclear component (Witschi and Aldridge, 1968). Silver and mercuric ions inhibit *in vitro* both the replication and transcription reactions by binding very strongly to the bases and forming rigid Ag^+–DNA and Hg^{2+}–DNA complexes. The irreversible binding of silver ions to the cell nucleus is seen by the black silver inclusion in tissues in the pathologic condition known as argyria. Divalent mercury and CH_3–Hg^+ form strong cross-links between DNA strands, preventing the unwinding and rewinding of DNA in biologic systems. Divalent nickel ions within a cell of high metabolic activity bind firmly to the nucleic acids and inhibit DNA-dependent RNA synthesis (Heath and Webb, 1967; Kasprzak and Sunderman, 1969). High concentrations of divalent metal ions are known to induce the mispairing of nucleotide bases (Eichorn et al., 1973). This mispairing could cause errors in the propagation of genetic information.

Osmium tetroxide *in vitro* reacts with DNA in neutral media. The binding is mainly to thymidine and is irreversible; however, the binding does not appear to alter the DNA structure. Similar reactions occur with denatured DNA, but not with double-stranded native DNA (Beer et al., 1966). The reactions of osmic acid with DNA *in vivo* have not been studied.

The firm binding of uranyl ion (UO_2^{2+}) to nucleic acids is useful in tissue staining for electron microscopy. *In vitro* studies revealed the binding of uranyl ion to the adenine N_3 atoms in AMP, ADP, and ATP at pH 4.5 (Feldman et al., 1967); at alkaline pH, a dinuclear complex is formed in which the UO_2^{2+} is bonded to the phosphate and to the ribose hydroxyl atoms. The UO_2–ATP system at pH 6.8 is a 2:2 sandwich-type chelate, in which the uranyl ion is bonded to the β and γ phosphate groups and the ribose hydroxyl group (Agarwal and Feldman, 1968). Within a cell, the uranyl could be bound to 3′ phosphate, 2′ OH, and N_3 of adenine of a ribonucleic acid molecule. These bindings may be involved in uranium toxicity at the molecular level in the cell.

In vitro studies revealed the binding of three atoms of gold to a single deoxynucleotide; the binding is rapid with *d*-AMP or *d*-GMP, but slow with pyrimidines, and *d*-TMP binds only one atom of gold. Gold is presumed to bind to primary amines and the heterocyclic nitrogen atoms in the base moieties, but not to the phosphate groups, because Au-*d*-AMP, Au-*d*-ADP, and Au-*d*-ATP complexes are found to be insoluble in water, whereas the potassium salts of these gold deoxynucleotides are water-soluble. Gold could bind to these nitrogen atoms of DNA if gold ions found their way into living cells and disrupted the normal functioning of DNA.

For optimal activity *in vitro*, RNAse from lysosomal and other sources needs a certain concentration of divalent metal ions such as Mn^{2+} or Cu^{2+} *In vitro* studies have shown that binding of metals such as Ag^+, Hg^{2+}, or In^{3+} to RNA prevents the RNA degradation by RNAse (Singer and Frankel-Conrat, 1962). During metal intoxication by these metals in living cells and tissues, the lysosomal RNAse function could be drastically inhibited.

DNA molecules carry all the genetic information of an organism throughout its life span, and different portions of this information are released when required. The stabilization of the double helix of DNA in the cell is due to the binding of DNA with lysine- and arginine-rich proteins to form nucleoprotein; the protein part of the nucleoprotein presumably blocks the expression of the DNA code and the transcript of mRNA. The positively charged nuclear proteins are bound to the phosphate groups of DNA. Metal ions compete with these proteins for the phosphate groups of DNA; they can cleave the proteins from the nucleoproteins and liberate DNA for its normal functioning whenever it is required by the cell or the organism. Toxic metal ions could influence this cleavage in different ways to permit transcription of various segments of DNA, leading to abnormal cellular differentiation or synthesis of abnormal proteins.

Ribosomes are another type of nucleoproteins in which a nucleic acid chain is tightly bound to proteins; divalent ions are required for this tight packing or binding (Gavrilova et al., 1966; Goldberg, 1966). The metal ions are associated with ribosomes through ribosomal RNA. The ribosomes consist of particles 30 S and 50 S as designated by their sedimentation behavior in the ultracentrifuge. In the presence of Mg^{2+} at the 10^3 M level, the 30 S and 50 S particles associate to form the 70 S particle; at the 10^{-2} M Mg^{2+} level, they form the larger 100 S particle; below the 10^{-4} M Mg^{2+} level, there is no association. Removal of metal ions from ribosomes causes dissociation into smaller components and also the unwinding of smaller fragments. Addition of essential trace metal ions such as Zn^{2+} and Ni^{2+} rewinds the ribosomal fractions in a manner similar to the unwinding and rewinding of DNA. The binding of tRNA to the ribosome also requires divalent metal ions. During experimental intoxication in rats with intravenously administered Ce^{3+} and other lanthanide salts, there was considerable dissociation of ribosomes of the endoplasmic reticulum of the liver (Arvela and Karki, 1970). Other toxic metals, especially polyvalent cations, could behave in a similar manner.

The influence or the effect of toxic metals in the many aspects of protein biosynthesis is highly complex. The mechanism of metal toxicity at subcellular levels could be partly attributed to the irreversible binding of the metals to the nucleic acids.

BIOLOGIC MEMBRANE–METAL INTERACTIONS

A thin, fragile lipoprotein membrane about 50–100nm thick regulates the flow of material between cell cytoplasm and the outer environment. The membrane has high electrical impedance, low surface tension, and high lipid content. Davson and Danielli (1952) postulated that a double layer of phospholipid sandwiched between two layers of proteins constitutes a simple conventional biologic cell membrane. Protein islands interspersed in the sandwich may protrude above, below, and through the basic membrane. Organelles present in the cytoplasm have similar membranes surrounding them; however, mitochondria, nuclei, and chloroplasts are surrounded by double membranes. The membrane is stabilized by: (1) ionic bonds between phosphate of the central phospholipid layer and cations of lysine or arginine residues of the protein layer, and (2) Van der Waals bonds between lipid chains and hydrophobic ends of neutral amino acids of the protein and between opposing or neighboring lipids. Other cells, such as erythrocytes in mammalian systems, differ somewhat from this above conventional membrane. Irrespective of microscopic morphology, the chemical makeup is such that all these plasma membranes have net negative charges. Some cells may have net positive charges that are related to the cell wall components. Some cell membranes contain Ca^{2+} and Mg^{2+} ions integrated into them. When the Mg^{2+} ions are removed from a bacterial cell membrane, the membrane falls apart. Similarly, the stability of the isolated bovine erythrocyte membrane depends on the presence of Ca^{2+} and Mg^{2+} ions. The permeability of the plasma membrane is maintained by electrostatic forces. Toxic metal ions may interact electrostatically with the cell membrane, especially with the phospholipid fraction; Cu^{2+}, Ag^+, and Hg^{2+} would bind firmly with the thio ligands in the protein fraction. The resultant conformational changes would have a major effect on the functioning of these membranes.

In general, divalent cations tighten the membranes and decrease their permeability, whereas monovalent cations loosen membrane structure and tend to increase the permeability of the membranes. Freshly laid trout eggs become so toughened when soaked in Ca^{2+} solution that a 5-kg weight is required to crush them (Manery, 1966). Phosphoryl ligands on some plasma membranes are involved in sugar phosphorylation and transport. Uranyl ions have been shown *in vitro* to combine into reversible and stable complexes with the phosphoryl and carbonyl ligands on membrane surfaces and inhibit sugar phosphorylation and transport. Other toxic metal ions could behave similarly.

Following entry into the blood, metal ions approach the erythrocytes as prime targets, although other metal-binding ligands are in abundance in the blood plasma. Hence, the toxic effects of metals on erythrocyte and its membrane are readily explored. Polyvalent ion species affect the electrophoretic mobility, permeability, and agglutination of erythrocytes. Carboxyl groups of neuraminic acid, which is an integral part of erythrocyte membrane, are known to determine erythrocyte electrophoretic mobility. The polyvalent cations bind to the phospholipid ligands, which do not normally contribute to the surface charge, and this binding confers excess positive charges and changes the electrophoretic mobility of the erythrocytes. The biologic impact of this charge reversal in erythrocytes is not fully known. The binding capacity of metals toward the phosphate of phospholipid is: $Th > UO_2 > La > Cu; Ni > Ca;$ and $Sr > Ba > Mn$ (Eylar et al., 1962). Agglutination is caused by a phenomenon other than the charge-reversal effect; multivalent cations combine with the carboxyl groups on the erythrocyte surface, reduce the negative charges, and facilitate agglutination (Jandl and Simmons, 1957). The agglutinating capacity of the ions can be rated as follows: Ca^{2+}, Cu^{2+}, Th^{4+}, $Tl^{3+} > Al^{3+}$, Fe^{3+}, Cr^{3+}, Be^{2+}, Sn^{4+}, $Zn^{2+} > Rb$, $Ir^{4+} > Ag^+$, $HAuCl_4 > Cd^{2+}$, $Ni^{2+} > Mn^{2+}$, Co^{2+}, Hg. Hemolysis takes place readily with Hg^{2+}, Ag^+, Cd^{2+}, and $HAuCl_4$. Metal ions such as Cr^{3+}, Fe^{3+}, and Be^{2+} are capable of firmly attaching proteins to the erythrocyte surface. These metals "sensitize" erythrocytes to the agglutinating action of antibodies (Jandl and Simmons, 1957). Rabbit and dog erythrocytes treated with chromium chloride and human plasma proteins become sensitized against antihuman serum. Other factors are involved in erythrocyte agglutination. Agglutination occurs when the cells are put in a medium at pH 5.0 that is devoid of any metal ions. Trivalent chromium and iron are attached to the erythrocyte membrane, but do not enter the mature cell; lead and mercury ions penetrate the membrane easily; copper and zinc ions equilibrate slowly. Manganese enters the cell by simple diffusion, and Tl^+ accumulates in the cell through the sodium pump mechanism.

Mercury ions block glucose and glycerol transfer into erythrocytes; Cu^{2+} inhibits glycerol transfer; Au^{3+}, Hg^{2+}, and Pb^{2+} increase the passive alkali ion permeability. Inside the erythrocyte, complexing agents such as hemoglobin, glutathione, phosphoric acid esters, and functional –SH groups compete for the metal and may protect intracellular enzymes; Pb^{2+} causes a dramatic decrease of intracellular ATP. Mitochondrial membranes also bind toxic polyvalent metal ions to affect their permeability. Lanthanides bind to active phosphate groups and very effectively inhibit Ca^{2+}, Sr^{2+}, and Mn^{2+} transport across isolated rat liver mitochondria (Mela, 1968). Similar phenomena could occur in vivo during lanthanide intoxication.

Lysosomal membranes are susceptible to lysis by polyvalent metal ions; the rupture releases active acid hydrolases and may destroy this and neighbor cells.

5

CARCINOGENICITY AND TERATOGENICITY

CARCINOGENICITY

Tumors are abnormal masses of tissue that grow and persist independently of surrounding structures. They have no physiologic function. Tumors that spread to other tissues (metastasize), or are transplantable to other tissues, are called *malignant* tumors, or *cancers*. Tumors that are usually encapsulated and do not metastasize are *benign*. Since benign tumors may develop into malignant tumors, the U.S. Environmental Protection Agency (Gibney, 1975) classified as *oncogenic* those substances capable of inducing either type of tumor. The relationships of tumors that are involved with mineral toxicities are presented in Table 5-1.

Most malignant tumors have compound names and are complex manifestations of many pathologic conditions. Cancer is malignant cellular growth with multiplication or survival that exhibits relative autonomy from control mechanisms of the host organism. The prime characteristic of cancer is metastasis with infiltrative and destructive cell proliferation. Carcinogens, or oncogens, are biologic, chemical, and physical agents capable of producing uncontrolled cell proliferation in organs and tissues. Carcinogenicity varies with different routes of administration, exposure time, dose, and physical state of the material, as well as with host-specific factors.

Carcinogenic agents can be divided into four main categories: (1) ionizing radiation, radioactive metals, and ultraviolet light; (2) oncogenic viruses; (3) chemical agents—nonradioactive materials, which include organic compounds of molecular weight below 500 gm, metals, and inorganic compounds; and (4) nonspecific surface oncogens. The last

TABLE 5-1. Relationships of Tumors Involved with Mineral Toxicities

Source	Type	Examples
Epithelial tumors	Benign	papilloma
		adenoma
	Malignant	papillary carcinoma
		adenocarcinoma
Mesenchymal tumors	Benign	fibroma
		myoma
		osteoma
	Malignant	fibrosarcoma
		myosarcoma
		osteosarcoma

includes a wide variety of chemically inert polymeric materials and metals in the form of smooth foils, plates, microscopic spheres, or needles. Chemicals, viruses, and radiation are the major primary causative agents of cancer in mammals.

Metals and their compounds induce cancer after a long latent period during which the metals evidently undergo biotransformation to initiate irreversible cellular changes necessary for the subsequent cancerous development. Metals may enhance tumor formation in tissue where initiation has already occurred. The interrelationships among metals in potentiating or enhancing carcinogenic effects are complex. It should be noted that the carcinogenic effects of organic carcinogens, such as azo dyes and diethylnitrosoamine, are additive and synergistic; some metals also act in these capacities. These effects are noticed between noncarcinogens and a single carcinogen; an otherwise innocuous environmental pollutant will act synergistically with an environmental carcinogen.

Cairns (1975) suggests that most cancers may be caused by exposure to environmental pollutants. Most metals occur in the environment in doses too low to cause immediate concern. Rarely are humans exposed to a low dose of a single chemical in an otherwise pure environment. Although naturally occurring environmental chemicals are known to induce different types of cancer, industrial operations contribute the major share of environmental carcinogens because many dusts and chemicals become concentrated in urban and industrial environments. However, some environmental carcinogens that cause cancer through unusual routes of exposure, such as intraperitoneal, intraosseous, or intratesticular, should not be indicted to the same degree as those that induce cancer following ingestion or inhalation. Urban air, food, and water are also major sources of carcinogens

(Fishbein, 1975). Although organic carcinogens now receive the attention of medical and industrial investigators, inorganic carcinogens need increased consideration (Saffiotti and Wagoner, 1976).

Except for radiation, no specific atomic physicochemical properties associated with the carcinogenicity of a metal have been experimentally established; however, general concepts correlating carcinogenicity with certain physicochemical properties can be mentioned. The electronegativity, solubility, and particle size of the metal and its compounds influence the carcinogenicity of the metal. Increased electronegativity or decreased electropositivity appears to be related to carcinogenicity. A survey reveals that carcinogenic metals possess electronegativity ranging from 1.2 to 1.9 (Table 5-2). These metals form covalent and coordination compounds with biochemically important ligands. Metals with electronegativity below 1.2 form electrovalent compounds and ionize readily. Most metals with electronegativity greater than 1.9 are too heavy to be reactive; amphoteric selenium is the obvious exception. Of Group V metals, arsenic may be chemically carcinogenic. Other metals in the midrange of electronegativity, namely thallium, indium, and mercury, are too toxic to have been tested for chemical carcinogenicity.

The solubility of the salts of some carcinogenic metals is inversely proportional to their tumorigenic properties; e.g., oxides and sulfides of

TABLE 5-2. Electronegativity of Carcinogenic Metals

Electronegativity	Chemical carcinogenicity		
	Positive	Suspected	None
0.7			Cs
0.8			K,Ru
0.9			Na,Ba
1.0			Li,Ca,Sr
1.1			La
1.2		Y	Ha
1.3		Sc	
1.4	Zr		
1.5	Be,Cr	Al,Ti	
1.6	Zn	Ga	V
1.7	Cd		W,In
1.8	Co,Ni,Sn,Pb	Mn,Fe,Si	Mo,Tl,Ge
1.9		Cu	Tc,Re,Hg,Sb,Bi
2.0		As	
2.1			Te
2.2		Rh	Ru,Os,Ir,Pt
2.4	Se		Ag,Au

nickel and chromium are carcinogenic (Payne, 1964; Gilman, 1966). Readily soluble salts may cause extensive necrosis at the site of injection before absorption, while sparingly soluble salts remain at the site and diffuse very slowly as they are phagocytized and undergo transformation to cause tumors. This relationship cannot be extended to all soluble salts, however, because chromates, selenates, and some other salts cause cancer following chronic ingestion at low doses.

Many carcinogenic metals are essential nutrients; they induce acute toxicity at high concentrations. The carcinogenic effect frequently occurs at lower, chemically nontoxic levels, and is noted a long time after the initial exposure. This time interval may be 6–20 months in mice (Schroeder, 1975) and 10–20 years in humans; Cairns (1975) cites epidemiologic evidence showing the rise in lung cancer 20 years following increased cigarette smoking in different segments of our population. Sometimes a dose is far below the acute or chronic LD_{50} dose levels of, for example, nickel or chromium. It is difficult to establish the lowest dose at which a metal can be carcinogenic. The accumulation of minute quantities of radioactive metals during lifetime studies indicates one difficulty in establishing a cause–effect relationship in uncontrolled environments. Little *et al.* (1975) reported that lung cancer was obtained experimentally in hamsters with 9–15 rads of radiation from ^{210}Po, equivalent to that found in the smoke from two packs of cigarettes per day for 25 years. Irritation from such low-level radiation may be synergistic to other carcinogens or cause local reactions, making the lung more susceptible to infection. The chemical toxicity of a metal differs from its carcinogenicity. At high doses, chemical toxicity kills the cells; at sublethal doses, the cells can recover and may acquire increased resistance to chemical toxicity. Carcinogens cause permanent damage and a biologic modification of the cells, making them more susceptible to further carcinogenic action. Some metals are primary carcinogens; others are cocarcinogens. Metal cocarcinogens may enhance or potentiate the carcinogenicity of organic compounds by inhibiting detoxicating mechanisms; e.g., copper potentiates the tumorigenesis of N-hydroxy 2-acetylaminofluorene (Stanton, 1967). Copper also accelerates the development of mouse skin tumors induced by benzanthracene compounds (Fare, 1965). Metals such as beryllium, chromium, cobalt, nickel, cadmium, selenium, and zinc are proven carcinogens. Special organic complexes of titanium, iron, nickel, rhodium, and palladium are established carcinogens, e.g., titanocene and iron dextran. Scandium, arsenic, selenium, yttrium, zirconium, silver, and lead are reported to possess carcinogenicity, while manganese, copper, zinc, and tin may possess recondite carcinogenicity. Mercury, niobium, tellurium, tantalum, and gold are reported to be nonspecific, surface carcinogens. Silicon, magnesium, and iron (in asbestos) are also surface carcinogens. Metals and their compounds rarely induce skin cancer. Some minerals induce cancer at the site of primary contact, e.g., nickel, chromium, or asbestos in the lung or nasal sinuses. Most carcinogenic metals induce cancer at the site of secondary storage or retention, e.g., thorium, nickel, cobalt, or gadolinium in bone or bone marrow.

The role of some metals and their compounds in carcinogenesis was recognized in 1942 by Schinz and Uchlinger, who discussed the evidence for the carcinogenic activity of arsenic, cobalt, and chromium. The carcinogenic activity of several metals is well established. Inorganic carcinogens can be classified into three groups: (1) man-made or naturally occurring radioactive metals such as plutonium, polonium, and actinides (including different isotopes and salts of these elements); (2) nonradioactive metals and their compounds; (3) metal surface oncogens, and (4) mixtures of inorganic with organic carcinogens. While radioactive metals can induce cancer following a single intense exposure, other groups of agents usually induce cancer following chronic exposure. Carcinogenic metals, selected mixtures, and surface oncogenesis will be discussed in this chapter. If the only evidence that a metal is carcinogenic is from nonspecific surface oncogensis, that metal is not considered to be inherently carcinogenic in our evaluation. Metals with significant radioactivity are not discussed, because it is difficult to separate chemical from radiation carcinogenicity. Potassium may have significant radiation. Vanadium, rubidium, indium, cerium, lutetium, samarium, platinum, lead, and other metals have such exceedingly long half-lives that they are biologically insignificant.

Carcinogenic Metals

The distribution of metals with inherent carcinogenicity in the periodic chart is shown in Table 5-3. Most metals in the fourth period are carcinogenic. It can be assumed that the carcinogenicity is related to the electronic structure of these transitional and inner transitional metals. Details and references are given in Volume 2 (Venugopal and Luckey, 1977).

Group I

The alkali metals are not chemical carcinogens, but the subgroup IB metals copper and silver exhibit carcinogenicity. Radiation from ^{40}K, about 20 mrads per year in humans, may be of significance in the aged.

Copper. Epidemiologic evidence, such as a high incidence of cancer among coppersmiths, suggests a primary carcinogenic role for copper (Agnese *et al.*, 1959); however, this role could not be firmly established in

TABLE 5-3. Metal Carcinogens

◯ Nonspecific oncogen ☐ Proven carcinogen ⌐⌐ Suspected with some evidence ⌐⌐ Special complexes are carcinogenic ── Radioactive

Period	IA	IIA	IIIB	IVB	VB	VIB	VIIB	VIII			IB	IIB	IIIA	IVA	VA	VIA	VIIA	O
1	H																	He
2	Li	[Be]											B	C	N	O	F	Ne
3	Na	(Mg)											[Al]	(Si)	P	S	[Se]	Cl	Ar
4	K	Ca	[Sc]	[Ti]	V	[Cr]	[Mn]	(Fe)	[Co]	[Ni]	[Cu]	[Zn]	[Ga]	Ge	[As]	Se	Br	Kr
5	Rb	Sr	[Y]	(Zr)	(Nb)	Mo	Tc	Ru	[Rh]	[Pd]	(Ag)	[Cd]	In	(Sn)	Sb	(Te)	I	Xe
6	Cs	Ba	La	Hf	(Ta)	W	Re	Os	Ir	Pt	(Au)	(Hg)	Tl	[Pb]	Bi	Po	At	Rn
7	Fr	Ra	Ac															

| | | | Ce | Pr | Nd | Pm | Sm | Eu | Gd | Tb | Dy | Ho | Er | Tm | Yb | Lu |
| | | | Th | Pa | U | Np | Pu | Am | Cm | Bk | Cf | Es | Fm | Md | No | Lr |

experimental studies with rats (Gilman and Ruckerbauer, 1962). This observation would suggest that impurities in the commerical metal may be involved. The cocarcinogenic character of copper is accepted. The significance of the increased concentrations of copper, zinc, and iron in malignant tumors induced by organic carcinogens is not clear (Arnold and Sasse, 1961). A higher incidence of stomach cancer in humans has been found in regions where the Zn:Cu ratio in the soil exceeded certain limits (Stocks and Davies, 1960). Stanton (1967) found organic copper complexes produced tumors in rat bone and lung.

Silver. Silver foil has induced fibrosarcomas in rats at the site of implantation, especially subcutaneous implantation (Oppenheimer *et al.,* 1956). Intravenous injection of colloidal silver also induced tumors in rats (Schmahl and Steinhoff, 1960). Both these findings may reflect the nonspecific action of surface oncogenesis.

Gold. Collodial gold does not induce carcinogenesis in rodents, but the nonspecific tumor formation associated with smooth surfaces is noticed following subcutaneous implantation of gold foil.

Group II

Among the subgroup IIA metals, only beryllium shows chemical carcinogenicity. Subgroup IIB metals are carcinogenic.

Beryllium. The incidence of cancer among human beryllosis victims implicates beryllium as an occupational carcinogen, but the incidence of cancer in exposed workers is less than 2% (Stockinger, 1966). The carcinogenicity of beryllium is well established in experimental animals. Primary lung tumors develop in rats and monkeys following chronic inhalation of aerosols containing beryllium sulfate or oxide (Schepers *et al.,* 1957; Vorwald *et al.,* 1966). Osteogenic sarcomas are induced in rabbits following intravenous injection of soluble beryllium salts. The topical deposition of beryllium into the medullary cavity of long bones in rodents induced malignant tumors (Tapp, 1966). Potential mechanisms for the carcinogenicity of beryllium are established by the high and selective affinity of beryllium for the cell nucleus, by the *in vitro* inhibition of enzymes such as Mg^{2+}-activated RNA polymerase and deoxythimidine kinase by beryllium, and by the interference of beryllium *in vitro* with differentiation of developing embryonic tissue.

Magnesium. No studies indicate that magnesium is a specific carcinogen. Its occurrence in asbestos, as magnesium silicate, involves it in nonspecific surface oncogenesis. As noted previously, thymic tumors are a notable symptom of magnesium deficiency.

Zinc. Evidence for zinc involvement in mammalian cancer is increasing. In addition to the nonspecific surface oncogenesis caused by the parenteral implantation of zinc foil, intratesticular injections of zinc salts into rats are reported to produce tumors (Riviere *et al.*, 1960). Mammary carcinomas have been reported in rodents following the addition of zinc chloride to their drinking water (Halme, 1961). Zinc is found at an elevated concentration in mammary tumors induced by other carcinogens (Tupper *et al.*, 1955). The cocarcinogenesis of zinc is established by the enhanced growth of experimental sarcomas induced by aromatic polynuclear hydrocarbons (Chahovitch, 1955).

Cadmium. The primary carcinogenicity of cadmium is well established. Several cadmium compounds have been reported to be carcinogenic in experimental animals and birds. Cadmium and its salts, however, were not found to be carcinogenic when administered orally to experimental animals (Sunderman, 1971). Soluble cadmium chloride at 5 ppm of cadmium in the drinking water of mice did not cause carcinogenesis during chronic exposure of 1 year (Schroeder *et al.*, 1964). The subcutaneous implantation in rats of cadmium foil or the intramuscular injection in rats of cadmium powder suspended in fowl serum caused transplantable rhabdomyosarcomas to develop at the sites of injection (Heath *et al.*, 1962; Heath and Webb, 1967). The insoluble oxides and sulfides of cadmium caused spindle cell fibrosarcoma in rats 6 months after intramuscular injection (Kazantezis and Hanburg, 1966); subcutaneous injection of cadmium oxide suspension produced tumor *in situ* (Malcolm, 1972). Soluble cadmium sulfate and cadmium chloride induced sarcomas in rats and mice at the site of injection when given parenterally and in the testes when given subcutaneously or intravenously (Haddow *et al.*, 1964a). Intratesticular injection of cadmium chloride caused tumors *in situ* in rats (Haddow *et al.*, 1964a) and in fowl (Guthrie, 1964). The cadmium content of malignant tumors induced by organic carcinogens increased from trace levels to about 26 mg per kg wet tissue (Gorodiskii *et al.*, 1956). There is no substantive evidence for cadmium carcinogenesis in humans; however, high concentrations of this metal in some cancerous tissues of human cadavers from cadmium intoxication implicates cadmium in human cancer (Butt, 1960). Gurr *et al.* (1963) found zinc prevented interstitial tumors induced by cadmium.

Mercury. Nonspecific surface carcinogenicity is exhibited by fine droplets of mercury injected into rats intraperitoneally for 2 weeks. Spindle-shaped sarcomas developed in abdominal muscles following a latent period of nearly 2 years; these sarcomas contained fine droplets of mercury (Druckery *et al.*, 1957). Absence of reports on specific mercury salts suggests that it is not inherently carcinogenic. It has long been used as a medicinal.

Group III

Among Group III metals, the actinides are carcinogenic due to their high radioactivity. There are unconfirmed reports of the carcinogenic effect of scandium, yttrium, aluminum, and gallium. Surface oncogenesis has been reported with several Group III metals.

Earlier observations that aluminum is carcinogenic had no valid experimental evidence. Cancer attributed to the use of aluminum cookware could have been due to nickel and chromium impurities in the aluminum ware. Aluminum foil induced myosarcomas in rats following subcutaneous and intramuscular implantations (O'Gara and Brown, 1967). Aluminum dextran was reported to induce sarcomas in rats (Haddow and Horning, 1960).

Soluble salts of gallium, scandium, or yttrium, when fed to mice at 5 ppm in drinking water over a long period of time, are reported to induce malignant neoplasms in the liver (Schroeder and Mitchner, 1971); these reports are yet to be confirmed. Subcutaneous implantation of metallic yttrium, cerium, praseodymium, gadolinium, dysprosium, and ytterbium in mice induced growth of granulomatous neoplasms (Talbot *et al.*, 1965). This carcinogenic effect was attributed to the "uncovering of latent oncogenic factors" (Ball *et al.*, 1970), and may be nonspecific surface oncogenesis. Osteosarcomas and other cancers produced experimentally by radioactive metals of the actinide series are highlighted by what is known about plutonium (Bair and Thompson, 1974).

Group IV

Metals of Group IV, with the exception of germanium and hafnium, may be considered to be suspected carcinogens; however, there are no substantive reports of their carcinogenicity, except a few that noted nonspecific surface oncogenesis. The carcinogenicity of silica in experimental animals is generally attributed to the polymeric nature of silica, impurities in the material, and nonspecific surface oncogenesis.

Silica, Silicon, and Asbestos. Evidence is accumulating to establish an essential nutrient role for silicon. Silica contribute to the formation of bone and cartilage (Hopps *et al.*, 1976). Opalphytolith, a polymeric form of silica found in the lymph nodes of sheep, led to the suggestion that phytoliths may have a role in carcinogenesis. Animal studies show that ingested asbestos is poorly absorbed; inhaled asbestos can be found in tissues such as the diaphragm, the liver, and tissues associated with the gastrointestinal tract. Asbestos bodies are not present in tumors of lungs, but are found in alveolar macrophages (Laguillaumie *et al.*, 1962). Materials resembling asbestos fibers in size and shape, such as taconite, also cause mesothelioma among mine workers. Epidemiologic evidence strongly implicates asbestos in the high incidence of cancer (mesothelioma) among asbestos miners and men working with insulation material. Human exposure to asbestos has become general, since drinking water, beverages, and pharmaceuticals filtered through asbestos fibers now contain minute quantities of the small fibers. Car brakes and construction spackling materials increase exposure for many people.

The etiology of asbestosis in carcinoma of the lung was recognized in 1935 (Lynch and Smith, 1935; Gloyne and Wood, 1935). There is an extended time period, sometimes 20 years, between the initial exposure and evidence of biologic effect, e.g., mesothelioma, the asbestos-induced cancer of the lining of the abdomen and chest in industrial workers. Pulmonary tumors were noted in mice exposed to asbestos dust (Lynch et al., 1957). Tumors were induced experimentally in fowls following the injection of asbestos particles (Peacock and Peacock, 1965).

Asbestos and silica are naturally occurring polymeric minerals that are implicated in cancer; asbestos can also cause severe physical damage to the lung by irritation. Asbestos encompasses a whole group of minerals that are distinct in chemical composition, crystal structure, particle size, and biologic activity; it may be one of the several silicate materials existing in either fibrous, or "asbestos," or nonfibrous form. Asbestos fibers are made of chains of SiO_2 bound together by iron or magnesium ions; other metals such as nickel, chromium, or cadmium are also present. Chrysolite (contains more magnesium than silicon) and crocidolite (contains more iron that silicon) are the two major fibrous forms of asbestos. Chrysolite exists in small bundles of very small hollow fibrils, usually less than 1 μm in diameter and 5 μm in length. These fine silicate fibers have a large surface area for absorption of trace metal ions. Crocidolite consists of solid bundles of fibers with large diameters and contains sodium, calcium, magnesium, and random trace metal ions in the fiber lattice structure. Both forms are associated with other minerals, and both are implicated in cancer.

The observations of Shin and Firminger (1973) and Zeedijk (1973) indicate the complexity of asbestos carcinogenicity. It is difficult to dissociate the physical from the electrical or chemical properties of asbestos in determining the causative carcinogenic agent. Since asbestos contains trace amounts of carcinogenic metals such as chromium and nickel, it is difficult to pinpoint a single carcinogen in asbestos (Dixon et al., 1969). Gross and Harley (1973) suggest that the metal components are not critical. The carcinogenicity of asbestos is attributable neither to its major components nor to any specific absorbed toxic metal. Asbestos carcinogenicity is attributed primarily to its unreactive polymeric nature, smooth surface, and particle size. As indicated under the heading *Surface Oncogenesis* (p. 148) silica and asbestos are surface oncogens.

Tin. Metallic tin exhibits nonspecific surface oncogenesis following implantation of small discs of smooth tin. Soluble polymeric salts of tin, such as sodium chlorostannate, induce sarcomas in rats, especially in the liver, following dietary ingestion (Walters and Roe, 1965; Roe *et al.*, 1965). Cationic tin salts in rat diets did not cause tumor formation.

Lead. Soluble lead salts are not generally carcinogenic; feeding 5 ppm lead in the drinking water of rats did not induce tumors (Schroeder *et al.*, 1964). Basic lead acetate, however, was found to induce malignant neoplasms in the kidneys of rats following dietary ingestion (Boyland *et al.*, 1962; van Esch *et al.*, 1962, 1969; Coogan, 1973). Mao and Molner (1967) give the tumor histology. Parenteral administration of insoluble lead phosphate produced renal adenomas and carcinomas in rats (Zollinger, 1953). Reports about cerebral tumors caused by lead intoxication among lead industry workers are not substantive. Lead powder is not carcinogenic either by inhalation or implantation. Tetraethyl lead is not known to cause cancer directly, but the breakdown products can cause cancer. Feeding lead arsenate did not induce cancer in rats (Kroes *et al.*, 1974).

Titanium. Soluble titanium salts are not carcinogenic *per os*. Titanium dioxide, suspended in trioctanoin, induced fibrosarcomas when injected intramuscularly into rats; titanocene (titanium complexed with dichlorodicyclopentadiene) induced a variety of malignant neoplasms at the site of injection in both the liver and the spleen (Furst and Haro, 1969*a*). It is not definitely known whether the carcinogenicity was due to the organic part of the molecule or to the titanium, or whether titanium potentiated the carcinogenicity of the organic compound.

Zirconium. Zirconium metal exhibits nonspecific surface oncogenicity following implantation in rat tissues. Orally ingested zirconium salts are not carcinogenic in experimental animals, but administration by parenteral or respiratory routes renders the zirconium salts carcinogenic (Sunderman, 1971).

Group V

Among Group V metals, arsenic and vanadium may be carcinogenic. Under unusual circumstances, tantalum and niobium exhibit nonspecific surface oncogenesis following implantation in animal tissues. There are no reports about the carcinogenesis of either antimony or bismuth.

Arsenic. The metalloid arsenic is a carcinogenic enigma. Earlier reports and reviews strongly assert the carcinogenecity of arsenic (Currie,

1947). There is a strong correlation between arsenic therapy and bronchial carcinoma in patients treated with arsenic compounds (Robson and Jelliffe, 1963). When treated with arsenic trioxide, asthma patients may develop Bowen's disease, which has clinical symptoms characterized by chronic precancerous dermatitis and multiple epitheliomatosis (Novey and Martel, 1969). Increased fatalities from respiratory cancers among workers in the arsenic smelting industry (Lee and Fraumeni, 1969) and the correlation between the increased use of arsenic sprays in agriculture and bronchial cancer among the spray handlers provide epidemiologic evidence for arsenic carcinogenesis (Roth, 1958; Blejer and Wagner, 1976).

The influence of sulfur dioxide or other unidentified chemicals that are associated with arsenic exposure was not taken into consideration in the Lee and Fraumeni studies. Pinto and Bennet (1963) commented that there is no evidence that chronic arsenic trioxide exposure caused systemic cancer among a group of smelter employees.

The toxic effects of arsenic in a British beer-poisoning episode correlated with the increased appearance of lung cancer among the beer drinkers (Satterlee, 1960). Skin cancer is prevalent in endemic areas of chronic arsenicism where drinking water contains 1–2.5 ppm arsenic. Rats and mice fed arsenic trioxide, at 34 mg/liter of water for 2 years, suffered from arsenic intoxication and not from arsenic carcinogenicity (Hueper and Payne, 1962). Arsenic carcinogenesis could not be produced in experimental animals (Shubik and Hartwell, 1957). In other experimental animals, arsenicals were reported to minimize the induction of cancer (Boutwell, 1963; Milner, 1969). Experimental and epidemiologic evidence of arsenic carcinogenicity is thus contradictory, although epidemiologic studies of industrial and farm workers in Europe strongly suggests that arsenic is a potent carcinogen.

Pentavalent arsenic acts as a cocarcinogen *in vitro* by suppressing host resistance factors (Stone, 1969) and helps in the survival of a spontaneously arising tumor; AsO_3^- appears to inhibit DNA repair mechanisms in *in vitro* experiments involving skin grafts (Jung and Trachsel, 1970). In comprehensive reviews on arsenic and its compounds, Frost (1967, 1970) suggests that traces of selenium impurities present in arsenates may be the toxic agent.

Vanadium. There is no direct substantive evidence of vanadium carcinogenesis. A statistical survey suggests the involvement of vanadium, arsenic, and zinc in human lung cancer (Stocks, 1960). A statistical correlation was established between lung cancer in human males and vanadium concentrations of ambient air. The capacity of inhaled vanadium trichloride to lower RNA and DNA contents of the liver, kidney, stomach, myocardium, and lungs of experimental animals may be involved in any cocarcinogenic action of vanadium (Roschin, 1967).

Group VI

Chromium and selenium are proven carcinogens among the metals of Group VI. There are no reports about the involvement of molybdenum and tungsten in carcinogenesis.

Chromium. Chromium carcinogenesis is documented by Baetjer (1956). Chromium is involved primarily in lung cancer, especially among workers in chromium metallurgy and refining and in dichromate manufacture from chromite ores. The cancer is not of any single histologic type; squamous cell carcinomas, round-cell carcinomas, and adenocarcinomas are reported. Pulmonary cancer is reportedly induced by chronic inhalation exposure. The time lag between the initial exposure and the onset of cancer ranges from 4 to 20 years; the average lag in Baetjer's study was 17 years. Chromium does not induce cancer in other areas of the human body, nor does it increase the incidence of cancer due to other carcinogenic agents. The actual carcinogenic chromium compound has not yet been identified; both Cr^{+6} and fine particles of insoluble chromite ore are considered to be the agents. Monochromate dust, fume, and mist have also been judged to be hazardous. A 1972 survey of workers exposed for about 20–25 years in a German plant involved in chromic oxide and chromic sulfate (Cr^{3+}) manufacture revealed no lung cancer among the workers (N.A.S., 1974). Thus, observations about the involvement of either Cr^{3+} or Cr^{6+} as the carcinogenic agent in human bronchial cancer are controversial. Chromium compounds appear to have no carcinogenic potential by the oral route. Chromate dusts and fumes are a great hazard, and chronic exposure increases the incidence of lung tumors (Schepers, 1971).

Earlier attempts to induce lung tumors in rodents with chromium and chromates were not successful (Baetjer *et al.*, 1959; Steffee and Baetjer, 1965). Intramuscular or subcutaneous injection of chromite suspension in lanolin into rodents induced bronchial cancer; before injection, the chromite was leached to remove soluble chromates (Payne, 1960). Subsequently, soluble and insoluble chromates were tested for carcinogenicity; soluble compounds had to be injected repeatedly to induce cancer. With intratracheal deposition, soluble chromates were not effective in inducing cancer, and the incidence was low (Hueper and Payne, 1962). It is now accepted that both Cr^{3+} and Cr^{6+} compounds are active in inducing bronchial carcinomas, with Cr^{3+} playing a key role.

The carcinogen in chromium-induced bronchiogenic cancer is Cr^{3+}, apparently in the form of insoluble Cr_2O_3 that has been inhaled chronically over a long period. Although the chromium content of the lungs was much greater than normal, the amount of either soluble or insoluble chromium bore no direct relationship to the incidence of cancer. A latent period seems to be necessary for the biotransformation of Cr^{3+} to a form that is carcinogenic.

Selenium. Selenium is a paradoxical element. There are no direct reports about selenium carcinogenicity in humans. Dietary selenium apparently induces cancer in rodents; the positive data override the negative.

Epidemiologic evidence suggests that selenium compounds may have therapeutic value against cancer in humans. The inverse relationship between ambient selenium level and human cancer mortality (Frost, 1972) does not support selenium involvement in causing human cancer. Synthetic lipids containing divalent selenium appear to prevent urinary excretion of –SH compounds, a result of lipid imbalance noticed in patients with cancer. The reproducible palliative effects of selenium support the hypothesis that selenium compounds may have therapeutic value against cancer.

Considerable experimental evidence supports the hypothesis that selenium induces cancer in rodents. Nelson *et al.* (1943) induced nonmetastasizing hepatic tumors in rats by feeding 5–10 ppm of selenium as seleniferous grain; the latent period for carcinogenesis was about 18 months. Feeding 4.3 ppm of selenium as the sulfate to rats induced cancer in 25% of them (Tscherkes *et al.,* 1961, 1963). Hepatic carcinomas, hepatic adenomas, and precancerous neoplasms were observed; the carcinomas metastasized to the lungs. Schroeder and Mitchner (1971) reported that 3 ppm of selenium as selenate in the drinking water of rats induced hepatic neoplasms following chronic exposure for 22 months; this low level of selenate also increased longevity in the animals.

Harr *et al.* (1967) reported that feeding 8–16 ppm of selenium in the diet to rats did not induce neoplasm; at these higher doses, the animals died due to the chemical toxicity of selenium before any tumors could develop. The negative results may also be attributable to the use of a different strain of rats. Selenium replaces sulfur, or competes with sulfur, in the formation of adenosyl methionine, which is involved in methylation reactions; selenomethionine is known to be present in proteins. Adenosyl-selenomethionine is known to cause abnormal methylation of nucleic acids *in vitro*. This abnormal methylation is yet to be directly confirmed *in vivo;* if confirmed, it could explain the carcinogenic activity of selenium.

Frost (1972) reviewed the evidence on the effect of selenium in the inhibition of experimentally induced cancer. Sodium selenite, 2.5 ppm selenium, delayed the formation of neoplasms in rats caused by a known carcinogenic level of N-2-fluorenyl acetamide (Harr *et al.,* 1972). Selenite apparently inhibited the induction of tumors and cancers by crude cycad or cycasin (Frost, 1972). Sodium selenide prevented the cocarcinogenic action of croton oil on the mast cell reaction of DMBA-initiated mouse skin tumors (Riley, 1969). Selenite, when put in anthelminthic drenches for sheep, reduced the incidence of ovine cancer (Wedderburn, 1972).

Tellurium. Although implanted tellurium foil is oncogenic, there are no reports about the carcinogenicity of tellurium; similarities in the physical, chemical, and inherent metabolic properties between selenium and tellurium lead to the speculation that tellurium could be a potential carcinogen.

Group VII

Among the metals of Group VII, only manganese is reported to have potential carcinogenic activity. Chronic subcutaneous and intraperitoneal injections of manganese chloride (0.1 ml of a 1% solution) induced increased frequency of lymphosarcomas over the controls in female mice. Mammary adenosarcomas and leukemias were also noted; the tumors were induced in tissues away from the injection sites (Paolo, 1964).

Group VIII

Among Group VIII metals, the essential trace metals cobalt and nickel are proven carcinogens, and special organic salts of iron, rhodium, and palladium are carcinogenic, while the others are not. Nickel is more potent than cobalt in inducing malignant tumors.

Iron. Epidemiologic evidence suggests that chronic inhalation of taconite or iron oxide ores (as hematite, Fe_2O_3) induces bronchial cancer (Dreyfus, 1936; Braum et al., 1960; Podhrazsky, 1957) and also gastric cancer (Kraus et al., 1957). Efforts to induce lung carcinomas experimentally in rodents with iron oxides were not successful; occasional benign tumors that did not metastasize were observed (Muller and Erhardt, 1956). Many other iron compounds were tested for carcinogenicity in experimental animals using dietary and parenteral modes of administration; ferric dextran was found to induce sarcoma in rats at the site of intramuscular injection (Richmond, 1957). Iron complexes found to possess no carcinogenic activity in rodents are iron saccharate (Kunz et al., 1963), iron sorbital (Fielding, 1962), and iron citrate (Lundin, 1961; Roe and Haddow, 1965; Braun and Kern, 1968). Iron ions and dextran are not carcinogenic when administered singly, but the iron dextran complex is carcinogenic in both rats and rabbits (Haddow and Horning, 1960; Haddow et al., 1964b). These tumors metastasized and were transplantable. Tumors developed irrespective of the sites of implantation. An insoluble iron compound containing the injected iron was formed at the site of injection. During the gradual disappearance of iron, there accumulated a mass of collagenous tissue that later became cancerous (Muir and Goldberg, 1961). Since iron is involved in RNA replication (Ivanov, 1965), perhaps this iron–dextran complex behaves as a charge transfer complex. Iron–dextran caused sarcoma in a human patient at the site of injection 3 years following iron–dextran therapy for anemia (Robinson *et al.*, 1960).

Iron oxides act as secondary carcinogens by enhancing the carcinogenicity of compounds such as benzapyrene; the iron oxides evidently serve as inert carriers to transfer the carcinogen in high concentrations to healthy cells (Stokinger and Coffin, 1968; Saffiotti *et al.*, 1968).

Cobalt. The carcinogenicity of cobalt is well established. Cobalt also acts as an inhibitor or suppressor of methylcholanthrene-induced tumors in albino mice (Kasirsky *et al.*, 1965). Dietary cobalt salts are not involved in carcinogenesis in mammals. Inhaled and parenterally injected cobalt compounds such as cobalt oxides and sulfide cause tumors in experimental animals. Cobalt metal has induced tumors at the site of implantation (Heath, 1956); it is not known whether this effect is due to the nonspecific surface oncogenesis or to the inherent carcinogenicity of cobalt. Tumors were induced in the thyroid glands of rats and at the site of injection of cobalt powder (Weaver *et al.*, 1956). In rabbits, parenteral injection of cobalt dust induced liposarcomas with hyperplasia of adipose tissue; the tumors metastasized and were transplantable (Thomas and Thiery, 1953). Epidemiologic evidence for the carcinogenicity of cobalt compounds among industrial workers is contradictory, but fibrosarcomas were induced in mice 6 months after the parenteral injection of an industrial refining dust containing cobalt oxide and cobalt sulfide (Gilman and Ruckenbauer, 1962).

Nickel. Nickel and its compounds are primary carcinogens. Epidemiologic and experimental evidence confirms the carcinogenicity of nickel in both man and animals. The carcinogenicity is manifested irrespective of the mode of administration and the nature of the nickel compounds. Different kinds of malignant tumors are induced at the sites of injection and in tissues away from the injection sites. Nickel metal in finely divided or in colloidal form, when administered parenterally to rats and rabbits, induced sarcomas of bone, muscle, nerve, and connective tissue. The frequency, rapidity, and ease with which nickel and its compounds induce sarcomas establish nickel as a primary carcinogen. Epidemiologic evidence suggests that inhaled nickel ore powder caused lung cancer among nickel refinery workers; some cocarcinogenic substances, such as arsenic salts, could have contributed to this carcinogenicity (Doll, 1958; Morgan, 1958).

Nickel powder induced lung cancer in rats and guinea pigs 18 months following inhalation (Hueper, 1958). Tumors were produced in rats by subcutaneous, intramuscular, or intrafemoral injections of nickel powder (Hueper, 1955; Heath and Daniel, 1964). The tumors metastasized and were transplantable. A latent period of 4–5 months elapsed before the development of malignant tumors. Nickel sulfide induces cancer in strains of rats having different degrees of susceptibility; implanted disks, chips, and inhaled powder all induced sarcomas (Gilman and Herchen, 1963; Daniel, 1966). Methandrostenolone enhanced the carcinogenicity of nickel or its sulfide (Jasmin, 1963).

A single intramuscular injection of nickelocene (nickel complexed with dicyclopentadiene) suspended in trioctanoin produced transplantable tumors in rats (Haro et al., 1968). Inhalation of nickel carbonyl gas, an important intermediary in the commercial purification of nickel, was found to produce metastasizing lung tumors in rats (Sunderman and Donnelly, 1965). Nickel carcinogenicity from tobacco smoke is attributed to nickel carbonyl (Sunderman and Sunderman, 1961), which can inhibit certain enzymes (Sunderman, 1967).

The mechanism involved in nickel carcinogenesis is not known; nickel binds firmly to nucleic acid and inhibits DNA-dependent RNA synthesis (Heath and Webb, 1967; Kasprazak and Sunderman, 1969). The inhibition of benzpyrene hydroxylase and the inhibition of the induction of the enzyme that metabolizes benzpyrene is another possible mechanism.

Rhodium and Palladium. Rhodium trichloride, when fed to mice at 5 ppm in the drinking water in life-term studies, induced tumors. The induced tumors included lung adenocarcinoma and lymphoma-leukemia (Schroeder and Mitchner, 1971).

Palladium chloride, 5 ppm palladium in the drinking water, induced malignant tumors in the lungs of mice (Schroeder and Mitchner, 1971).

Mechanisms of Metal Carcinogenesis

There is no established mechanism or mode of action to explain metal carcinogenesis. Some essential trace metals are carcinogenic. Some metals have inherent carcinogenicity; others undergo biologic transformation to become carcinogenic, as indicated by the long latent period between initial exposure and onset of carcinogenesis. One aspect of aging in mammals is the random accumulation of heavy metals in the body and their localization in specific tissues without any apparent function, sometimes without exhibiting any chemical toxicity. In addition to the phenomenon of nonspecific tumor induction associated with surface oncogenesis, some metals have an intrinsic property, associated with cancer, that has not yet been established.

The intrinsic carcinogenic capacity of a metal might be explained by its ability to displace an essential metal that is required for the functioning of a critical enzyme or nucleic acid, thereby altering the normal metabolism of the cell or its multiplication. Since most essential trace metals are carcinogenic, an imbalance of metal ions is a more probable explanation. The depletion concept may be acceptable for the reason that metals are often antagonistic to similar metals, and an excess of one trace metal could actually increase the requirement of another. What is obvious is that rather small amounts of certain metals can change cells to release them from the usual controls that would otherwise limit the reproductive increment of that clone. This change appears to result from a nonphysiologic ratio between the carcinogenic excess of a metal and one or more bioconstituents of the cell.

The complexes of metallocenes, such as titanocene and nickelocene, potentiate the carcinogenicity of metals by acting like aromatic hydrocarbons; the carcinogenic activity could be attributed to the higher degree of lipid solubility of the compound over the metal ion or the metal inorganic compound. Szent-Gyorgyi (1960) explained the carcinogenicity of polynuclear aromatic hydrocarbons by postulating that the hydrocarbons form charge-transfer complexes with local receptors donating electrons. When bound to tissues or active biologic macromolecules, carcinogenic metals may act as similar charge-transfer complexes. The carcinogenic metal does not seem to possess any specific polarizability or specific geometric structure. Miller (1970) likened carcinogenic metal ions to the ultimate carcinogens of organic molecules that are potent electrophilic reactants. These are relatively positive, or electrophilic, atoms that react with relatively negative, or nucleophilic, atoms. For example, the divalent forms of beryllium, nickel, cobalt, lead, and cadmium (Furst and Haro, 1969a; Sunderman, 1968; Clayson, 1962) would readily react with DNA, RNA, or protein by displacing weakly electrophilic H^+, by forming insoluble phosphates, or by reacting with guanine. These reactions could produce neoplasia by affecting cell mutations, virus activation, changes in genome expression, or selection of latent tumor cells.

The cell nucleus appears to be the main target of carcinogenic metals. In primary rhabdomyosarcomata induced by intramuscular implantation of powdered metallic nickel, cobalt, or cadmium, at least 50% of the inducing metal that accumulates in the nucleus of the tumor cell is bound by the nucleoli; the rest is equally distributed between the nuclear sap and chromatin (Webb *et al.*, 1972). Moreover, the metal appears to be contained in the first cells that trigger the growth of the neoplasm, but is not found in the other cells of the growing neoplasm. The carcinogenic metals are electrophilic in the ionic form: Be^{2+}, Cd^{2+}, Co^{2+}, Ni^{2+}. These metals form insoluble phosphates and could conceivably interact with the tertiary phosphates of the nucleic acids and change the configuration of the nucleic acids. These metals can also replace the magnesium needed to maintain the double-stranded structure of nucleic acids. In fact, carcinogenic nickel, chromium, and cadmium are found in RNA isolated from phylogenetically different sources. Replication of DNA with DNA polymerase and transcription of the DNA code into m-RNA with RNA polymerase are mediated by magnesium ions. Tumor development in the thymus of rats fed a magnesium- or manganese-deficient diet (Bois, 1964) suggests that some other metal, evidently tumorigenic, has replaced magnesium in the magnesium-dependent RNAse system. This possibility suggests that the ratio of magnesium to other ions may be critical in tumor formation. Beryllium, which is known to inhibit DNA synthesis in *in vitro* systems and which is similar to magnesium in its outer orbital structure, may compete with magnesium during replication to incorporate abnormal deoxynucleotides and lead to the formation of aberrant DNA. The abnormal genetic coding mechanism could give rise to an autonomous cancerous cell that defies normal cell regulation mechanisms. The involvement of metal during fission and recombination of DNA in the DNA replication process is not clearly understood. Ivanov (1965) suggested that Fe^{2+} is involved during the fission stage. When Fe^{2+} combines with the bases of DNA, it is oxidized to Fe^{3+}, which stabilizes the single-stranded DNA molecule. On recombination of the two strands, the Fe^{3+} reverts back to Fe^{2+}; the whole sequence is dynamic. Carcinogenic metals can disturb this dynamic equilibrium by combining with the bases in the single DNA strand and preventing normal base pairing and the normal formation of the double DNA strand.

Lysosomal proteases and DNAses can act in abnormal ways. Carcinogenic and insoluble nickel salts, such as nickel sulfide, may gain entry into the cell by the adsorption of metal protein complexes at the surface with subsequent endocytosis (Heath *et al.,* 1969). Lysosomal proteases digest the carrier protein; the intracellular breakdown of nickel sulfide leads to the formation of "active" Ni^{2+}, which interferes with DNA replication as theorized above. The metal protein complex may lyse the lysosomal membrane to release DNAse, which can degrade the DNA molecule; before the degradation is complete, or during the process itself when the DNA is partially cleaved, the DNAse can be inactivated by the lysosomal cathepsins or the proteases. The overall outcome could be a surviving cell with a truncated or abnormal DNA. If this cell were to multiply using an abnormal DNA, it could become an automonous cancerous cell, capable of duplication without normal cell-regulation mechanisms. Adenosyl-selenomethionine is known to cause abnormal methylation of nucleic acids *in vitro;* this effect is yet to be directly confirmed *in vivo.* The carcinogenicity of selenium could be due to this abnormal methylation.

Essential trace metals are involved in maintaining the integrity of structural, functional, and catalytic proteins. Carcinogenic metals can replace, compete with, or equilibrate with the essential metals in metal-activated enzyme systems to change the enzyme kinetics. Any of these reactions may result in either enhancing or decreasing the enzyme activity. Retardation of the activity of a group of related catabolic enzymes, or enhanced functioning of anabolic enzymes, in a target organ or tissue could result in a steady state with constant increase in tissue deposition or tumor. The benign tumor may eventually become malignant. Maintenance of the disulfide–thiol cycle or equilibrium ($2\ \text{–SH} \rightleftharpoons\ \text{–S–S–} + 2\ \text{H}$) is necessary in a healthy cell for its normal growth and metabolism (Harrington, 1967). Carcinogenic metals can disrupt this cycle and induce an overcompensatory response that will result in abnormal cell growth and tumors. Excess of an essential metal can also induce abnormal activity in some functions of a cell. Excess of essential zinc, when added to a medium for a fungus, stimulated RNA synthesis by a factor of 2.5 (Wegener and Romano, 1963); this kind of overcompensatory response is not reported for mammalian tissues.

The cocarcinogenic activity of metals appears to be due mainly to inhibition of detoxicating systems involved in degrading organic carcinogens; the effect is indirect.

A latent period of 2–24 months seems to be necessary for the biotransformation of a harmless essential trace metal into a carcinogenic form. Chromium induces bronchiogenic cancer following long chronic inhalation of insoluble chromium oxide. Parenterally administered nickel sulfide or cobalt sulfide gains entry into cells in a form attached to a protein carrier; the protein is later degraded by lysosomal proteases. The biotransformation of the insoluble nickel or cobalt sulfides to some active carcinogenic form of the metals, the nature and mechanism of which are unknown, appears to take place at this stage. Accumulation of colloidal metal compounds in the lysosomes of the reticuloendothelial system causes this kind of activation or transformation.

Stress also seems to influence the induction of tumors. Golberg (1971) believes that most chemical compounds can induce experimental cancer in tissues under sufficient stress, and Boylard (1968) correlates experimental with epidemiologic evidence for carcinogenic materials.

Surface Oncogenesis

Surface oncogenesis is the phenomenon in which smooth-surfaced, noninflammatory implants of unreactive materials induce cancer. This toxicity is a latent one that takes many months to develop. The physical attributes of foreign particles in tissues are important. The size, shape, flexibility, and surface character determine the carcinogenicity of materials that have no inherent chemical carcinogenicity. Particle size and surface area are less important than surface characteristics. Oncogenic implantations include metals, silica, glass, asbestos, and plastics with smooth surfaces in the form of sheets, microscopic spheres, fibers, and small globules. The less inflammatory a material, the greater its carcinogenicity (Brand et al., 1967). Bischoff and Bryson (1969) reviewed much experimental work, and Autian (1975) reviewed present concepts of surface oncogenesis with polymeric substances.

Subcutaneous injection of colloidal silver induced tumors in rats. Peritoneal injection of fine droplets of mercury causes abdominal cancer in rats. Powdered beryllium, nickel, and cobalt produce tumors. Inhaled finely powdered nickel and insoluble chromates readily induce sarcomas in the lung. This carcinogenic action is attributed to the particle size of the colloidal or aerosol suspension of active carcinogens. Powdered gold, tantalum, quartz, or vitallium is not oncogenic in rats, while implanted small sheets of gold, silver, steel, tantalum, or quartz induced tumors (Schmahl and Steinhoff, 1960; Oppenheimer et al., 1956; Nothdurft, 1961; Shubik and Hartwell, 1962). Goldhaber (1962) indicated that there was a minimum smoothness threshold; the smoother and the less inflammatory, the more oncogenic the material. The critical minimum area is larger for subcutaneous than for intrapleural implantations. The closer to the head metal sheets are implanted, the more tumors noted. Thin tin foil (about 2 μm thick) did not induce cancer, but a thicker foil was found to be carcinogenic (Alexander and Horning, 1958). The etiologic factors responsible for the induction of tumor by a smooth surface are not definitely known. The phenomenon is nonspecific; both metals and nonmetals exhibit this oncogenesis. A long latent period makes studies very difficult. The latent period is ½–2 years in mice and rats and 1–2 decades in man.

Cairns (1975) suggests that most cancer is environmentally induced and correlates increased lung cancer in different populations with increased cigarette consumption 20 years previously. Cancers currently being found are attributed to silicon infusions performed a decade previously for cosmetic purposes.

Heath et al. (1971) showed that wear particles from artificial joints worked in Ringer's solution were carcinogenic to rats. To date, no evidence has shown tumor induction from metal prostheses or sutures in humans.

Most of the tumors reported for silicon, silica, or asbestos are probably due to nonspecific oncogenesis. Roy-Chowdhury et al. (1973) and others have suggested that the trace metal contaminants of asbestos are responsible for chemical carcinogenesis; this concept is contradicted by experiments with asbestos containing different trace metals (Gross and Harley, 1973), and by much information on surface oncogenesis. Although asbestos pollution is of great concern, other microscopic fibers are abundant in our environment. Rohl et al. (1975) have measured the incidence of a dozen mineral particles that become excessive in home environments during spackling, patching, and taping of walls. These particles can be harmful due to the chemical reactions of the mineral particle itself, to chemical reactions from material adsorbed onto the microscopic particles, or to the physical form of the particle. Neither the chemical action of the basic metals that comprise the inorganic minerals (asbestos, quartz, feldspar, talc, calcite, kaolinite, mica, and dolomite) used in dry-wall construction and repair nor the compounds usually adsorbed onto them are of major significance in chemical carcinogenesis.

Plastic solids also cause tumors in rats and mice. Turner (1951) found sarcomas in rats with subcutaneously implanted bakelite disks. Observations by Oppenheimer et al. (1948) of tumors following cellulose film implantation have been confirmed. Nothdurft (1955) explored this phenomenon and concluded that discs were more oncogenic than powders, and that solid disks were more effective than disks with holes. Millipore filter implantation induced more tumors in mice when the pore size was 100 nm than when it was 450 nm (Goldhaber, 1962). Implantation of nylon, polyethylene, and dotriacontane sheets caused localized tumors in rodents, while the same amounts of powdered plastic induced none (Oppenheimer et al., 1958; Hueper, 1961; Bryson and Bischoff, 1967). Hueper (1961) found a given amount of polyethylene-produced tumors in proportion to the surface area, except that powders produced none. Electropositive films or filters carrying detergent were more oncogenic than anionic films or electronegative (detergent-free) filters (Carter et al., 1971). Selye et al. (1962) found that subcutaneous implantation of 3-cm glass tubes induced tumors in 87% of the rats in 1 year, while 1-cm tubes induced tumors in only 8%. Implantation of hard waxes (Shubik et al., 1962) or the development of hard surfaces in vivo, such as cholesterol (Autian, 1975), can induce tumors.

Nonspecific surface carcinogenesis is enigmatic. Heath (1960) postulated that particulate metals or metal compounds may, by a mechanical process of irritation, help in the slow and continuing breakdown of differentiated fibroblasts into new myoblasts that are gradually transformed into malignant tumors. The involvement of the metal in the further growth of the original or the metastasized neoplasm is minimal, since the transplanted tumors do not contain the initiating carcinogenic metal, and further growth of the neoplasm does not depend on the presence of the metal. It is also possible that a minute tumor formed at the site of injection or at the point of initial contact is transported by the lymphatic system to the target tissue or to the site of major tumor induction.

Nonspecific surface oncogenesis is beyond present rationalization. Powdered glass is not carcinogenic; minute particles can be phagocytized. Fiber dimensions determine the carcinogenicity of smooth glass. Bischoff and Bryson (1964) summarized early theories associated with surface oncogenesis: (1) modification of the cellular environment, which acts as a promoter favoring the development of tumors from spontaneously arising precancerous cells; (2) interference with immunologic mechanisms that would otherwise hold in check potentially cancerous or neoplastic cells that arise from time to time under normal conditions; and (3) activation of latent oncogenic RNA viruses (viral oncogens incorporated into host genome). The last concept, also suggested by Langvad (1964), is reinforced by the inhibition of chemical carcinogenesis in mice by viral vaccines (Whitmore and Hueber, 1972). M. Luckey (1974) proposed that the bacteria-sized smooth particles maintain microloci in which inhibition of cell-wall contact is disrupted. This disruption would allow those cells to continue ontogeny and reproduction under abnormal conditions, and could account for the nonspecific carcinogenicity of smooth surfaces.

Surface oncogenic materials are noninflammatory and have a prolonged latent granuloma prior to cancer induction. Implanted subcellular-sized particles and rough-surfaced materials do not produce cancers. Smooth surfaces immobilize attached cells as effectively as the tar baby immobilized Brer Rabbit. This attachment removes a portion of the cell membrane from contact inhibition and alters cell polarity, metabolism, orientation of organelles, mobility, and ontogeny. Such cellular stress could allow ready infection by transient viruses, activation of latent viruses, or could simply attract fibroblasts to form a granuloma. Reproduction of cells within this abnormal mass, the granuloma, would allow mutant cells to form and survive without alerting the search-and-destroy mechanism of the immune system. Some of the cells become cancerous. Attainment of a critical mass by the cancerous clone of çells would be relatively easy in this immune-deficient environment. Beyond this critical mass, the immune system could be overwhelmed into a refractory state.

Van Peenen (1966) described the action of particles that are retained after being phagocytized.

> Almost any industrial dust which contains irritant particles of a size small enough to reach the bronchioles may cause a reaction in the lung. In all of them, a granulomatous foreign body reaction occurs which proceeds to severe fibrosis with all its functional sequences. Silica particles are especially irritant because of the piezoelectric properties of the tiny crystals. Silica usually elicits a far more striking response than the other irritants. Asbestos bodies and perhaps silica

particles are quickly ingested by monocytes and giant cells when first inhaled. This appears to protect the lung and minimizes reaction at first. However, as time passes and the phagocytes break down, the particles are released again, are apparently not rephagocytosed readily and cause considerable fibrosis. It has also been suggested that while in the phagocytes the particles are coated with protein which acts as hapten and that, on liberation of the particles, the hapten coating may elicit an antigen–antibody response which contributes to the fibrosis. These mechanisms explain why pneumoconioses are often, after the first exposure to dust, so long in becoming clinically apparent and progress even when the patient is no longer exposed to the inciting agent.

This protein hapten might involve the formation of organometallic complexes (equivalent to iron dextran) that fit the electrophilic or free radical theories of oncogenesis supported by Miller (1970) and Szent-Gyorgyi *et al.* (1960), respectively.

Oppenheimer *et al.* (1958) and Brand *et al.* (1967) studied cell activity in surface oncogenesis with an appropriate series of disc implantations in rats. The implant first become encapsulated with granulomatous tissue. After the first few months, cellular activity decreased markedly. There was a gradual disappearance of capillaries and a decrease in fibroblast cells; these oriented parallel to the plane of the implant. After 6 months, fibroblastic activity reappeared in patches of cell proliferation. The attached cells showed variations in size with anaplastic nuclei and showed atypical proliferation. A few months later, premalignant cells developed and penetrated the capsule, often forming nodules of active cells. Malignancy was detected within 1 month following the detachment of the premalignant cells. Extirpation of the implant and capsule prior to this time prevented tumor formation. Karyotype studies indicated that only one clone of malignant cells formed on each implant. If part of the implant were transferred to another host, tumors appeared in both donor and recipient at the same time.

Tumor induction resides in cell changes or metabolites during the latent period. Attachment of cells to the surface allows permanent change in cell structure and organelle orientation, interference in cell exchange of materials, and immobilization of the implanted material. As the granuloma develops to wall off the noninflammatory implant, the partial isolation of attached cells becomes critical. Decreased oxygen and nutrient supply with increased concentration of catabolic wastes, including fragments of dead cells, creates the environment for survival of a clone of cells that has developed, via mutation or differentiation, unresponsiveness to reproductive control.

Autian (1975) reviewed theories of nonspecific surface oncogenesis. Tumor incidence and length of latent period vary with species, strain, and condition of the animal; size, form, and hardness of the implant; and smoothness of the surface of insoluble materials. A solid surface of chemically unreactive material is required. The critical size for films is 0.5 cm diameter, disks and spheres have not been sized, needles may be microscopic, and fibers appear to be less oncogenic. Tight adherence of cells for several months (in rodents) prevents the usual exchange of metabolites and leads to initiation of precancerous cells. The constant irritation is related to cell attachment to a smooth surface—not to the edge of a film or disk, not to rough surfaces, and not to powders. The decreased inflammatory response contributes to tumor induction. The absorption of carcinogenic chemicals is possible for plastics and metal particulates, but the increased surface area and increased potential for absorption with decreased particle size has little effect in surface oncogenesis where there is a definite size limitation. The free radical theory might account for the carcinogenic property of some plastics, but not for that of glass, mercury, metals, or minerals involved in surface oncogenesis. The electrical effect of implanted materials must be considered. A piezoelectric effect related to particulate motion could give a minute local alternating current. The variety of materials includes glass and plastics, and the lack of direct evidence gives no emphasis to this component. When electrical fields stimulate new growth, the reproducing cells conform to normal organogenesis (Smith, 1974).

TERATOGENICITY

Teratology is the study of abnormal physical development and malformation of embryos. Mammalian embryos are susceptible to environmental influence, especially chemicals, as exemplified by the high susceptibility of developing human embryos to thalidomide (McBride, 1961; Lenoz, 1962). Inorganic metal salts are also teratogenic.

Fertilization, implantation, organogenesis, histogenesis, and functional maturation constitute the developmental functions of the mammalian embryo from conception to birth. The final stages of organogenesis and the early stages of histogenesis overlap; functional maturation and histogenesis overlap extensively. The developmental span is conveniently organized into two periods: (1) the embryonic period, during which organogenesis takes place; and (2) the fetal period, during which histogenesis and functional maturation take place. The embryo is susceptible to teratogenic agents with different degrees of vulnerability at the different stages of embryo development. Differentiation, mobilization, and organization of cell and tissue groups into organs take place during organogenesis. Toward the end of organogenesis and early histogenesis, refined cellular and tissue changes convert primordial organs into specific, definitive organs. Teratogenic agents interfere with organogenesis and cause gross structural defects or malformations. Teratogenic agents are radiation; excess of organic and inorganic chemicals; dietary deficiencies of vitamins, amino acids, and essential trace metals; infection; aberrant temperature; endocrine imbalance; physical trauma; placental failure; and maternal or fetal metabolic imbalance, or both. These agents can be lethal as well as teratogenic. Lethality occurs mostly during the early stages of conception from fertilization through cleavage in blastocyst and early germ layer stages. Teratogenic effects are more frequent during the critical embryogenesis or organogenesis period than in the fetal or maturation periods.

Teratogens induce growth retardation or functional disturbances or both in developing organs during the fetal period, which comprises later stages of histogenesis and functional maturation. During organogenesis, the developing embryo is highly susceptible to teratogenesis; this susceptiblity decreases as organogenesis proceeds toward completion and the embryo becomes progressively more resistant to teratogenesis. Teratogens can cause mutation, mitotic interference, chromosomal aberrations, abnormal nucleic acid metabolism, abnormal membrane characteristics, electrolyte imbalance, enzyme inhibition, and a host of other malfunctions. The combined effect of one or more of these malfunctions in the developing embryo may result in impeded morphogenetic development and lead to abnormal tissue or organ development in the newborn young or to embryonic death and fetal resorption. In addition to the inherent influence of the teratogens, the susceptibility of the embryo depends on the period of its development, the genotype of the conceptus, and its ability to counteract or resist the teratogenic agents. The mechanism of induction of teratogenesis is complex and is poorly understood, and it is difficult to associate a specific tissue malformation with a known teratogen in mammals. However, it appears that in avians, especially chicken embryos, teratogens act in specific manners leading to malformations (Landauer and Sopher, 1970) from interference with localized metabolic needs.

In mammals the placental membrane is an effective barrier and prevents some potential teratogens from reaching their target sites within the embryo. Maternal homeostasis regulates the concentration of some toxic agents that can reach the placental membrane. The placental membrane cannot effectively prevent the transport of chemical agents that are similar in structure to essential nutrients needed for the developing embryo. These nutrients include essential macrometals and trace metals. Both nutrient and metabolic antagonists may cross the barrier and become teratogenic. Dramatic changes in essential nutrient concentration may also be teratogenic. The essential metal zinc appears to be teratogenic in hamsters, rats, and chickens in both deficiency and excess. Manganese deficiency causes abnormal skeletal development in rodents. Cell division and differentiation in the developing embryo involve periods of intense enzyme activity and protein synthesis. The enzymes participating in the embryonic development are susceptible to toxic levels of those metals that can permeate the placental barrier. The rate of passage of metals depends on the physical and chemical properties of the metal ions and their compounds. In addition to size and charge of the metal ions, the time required to reach the developing embryo is influenced by the lipid solubility, degree of ionization, possible complexation, or chelation with other chemical compounds, and concentration of the toxic metals.

Maternal homeostasis and detoxication mechanisms control the level of toxic metals that can reach the placental membrane. Thus, except in intrauterine injection, it is difficult to establish the levels for teratogenic agents for any species or strain of experimental mammals. The embryo lethality and teratogenicity associated with a toxicant usually occur at a small fraction of the maternal toxic or lethal levels (LD_5 to LD_{50}); also, teratogens may have a no-effect dose range. The entire sequence is depicted in Fig. 5-1.

The relationship between dose and effect of teratogens indicates the possibility that embryocidal and teratogenic effects may be manifested at about the same level or may overlap at a given dose. Both effects may increase at parallel rates with increased dosage. Similarly, the embryotoxic and maternal toxic levels may overlap. The ratio of embryo lethal

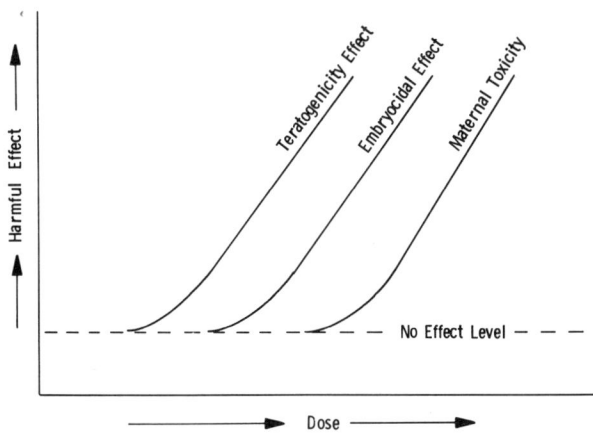

FIG. 5-1. Model of relationship between dose of a teratogen and effect on embryo and mother.

and maternal lethal dose is highly variable, depending on the physical condition and nurture of the maternal organism and the nature of the metal compound. The levels and teratogenic effects of essential metals follow different patterns, as depicted in Fig. 5-2.

It is more difficult to assess the embryocidal and teratogenic dose of an essential mineral under a deficiency state than it is under excess conditions; most of the essential minerals will be utilized by the dam rather than by the embryo.

Although purely speculative at the present time, it is possible that metals and their compounds can cause secondary or indirect teratogenesis. Metals and their salts cause a wide range of toxic effects in the pregnant mother. Derangement in maternal metabolism could lead to: (1) the formation of new toxicants that are not toxic to the mother but that are toxic to the developing embryo, and can permeate the placental membrane; and (2) the deficiency of an essential metabolite that cannot reach the embryo in adequate amounts and in proper time to ensure the programmed fetal development. The effects on the mother may harm the child *post partum* (Calley, 1974).

The distribution of teratogenic metals in the periodic chart is shown in Table 5-4. Some, such as copper and manganese, are not teratogenic, but are embryocidal; beryllium and aluminum are potential teratogens.

Among the Group I metals, lithium and copper have embryocidal effects. Lithium is also teratogenic. Lithium carbonate fed to pregnant

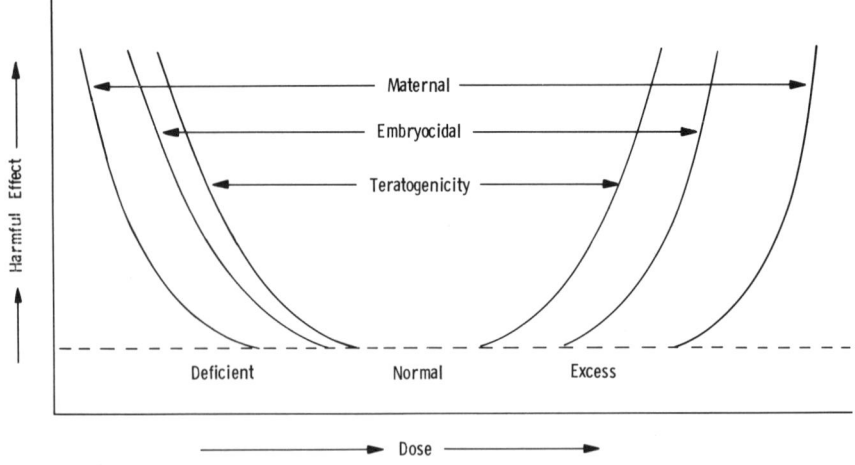

FIG. 5-2. Model of relationship between dose of an essential metal that is teratogenic and effect on embryo and mother.

CARCINOGENICITY AND TERATOGENICITY 157

TABLE 5-4. Teratogenic Metals

Legend:
- ─── Crosses placental barrier
- ⌐ Potential teratogen
- ☐ Teratogenic
- ○ Embryocidal

Period	IA	IIA	IIIB	IVB	VB	VIB	VIIB	VIII			IB	IIB	IIIA	IVA	VA	VIA	VIIA	O
1	H																	He
2	⊙Li⌐	⌐Be⌐											B	C	N	O	F	Ne
3	Na	Mg											⌐Al⌐	Si	P	☐As☐	☐Se☐	Cl	Ar
4	K	Ca	Sc	Ti	V	Cr	○Mn○	Fe	Co	Ni	Cu○	○Zn○	Ga	Ge			Br	Kr
5	Rb	Sr	Y	Zr	Nb	☐Mo☐	Tc	Ru	Rh	Pd	Ag	○Cd○	○In○	Sn	Sb	☐Te☐	I	Xe
6	Cs	Ba	La---Ce	Hf	Ta	W	Re	Os	Ir	Pt	Au	○Hg○	○Tl○	☐Pb☐	Bi	Po	At	Rn
7	Fr	Ra	Ac---Th		Pa		U	Np	Pu	Am	Cm	Bk						

Lanthanides: Pr, Nd, Pm, Sm, Eu, Gd, Tb, Dy, Ho, Er, Tm, Yb, Lu
Actinides: Pa, U, Np, Pu, Am, Cm, Bk, Cf, Es, Fm, Md, No, Lr

goats causes abortion with complete degeneration of the fetal liver; newborn kids that survive suffer from liver necrosis (Boulos et al., 1973). Pregnant rodents suffer embryotoxicity following oral ingestion or parenteral administrations of excess lithium chloride or oxide. Cleft palate and eye and external ear malformations are the teratogenic effects of dietary lithium. Szabo et al. (1970) established a possible threshold for lithium teratogenicity in mice at 200 mg lithium carbonate per kg mother per day, at 6–15 days of gestation. The occurrence of malformed babies born to mothers undergoing therapeutic treatment for maniac psychosis has been reported (Schou and Amdisen, 1970; Aoki and Ruedy, 1971). Lithium teratogenicity in experimental animals is dose-related and differs from species to species. Copper has an embryocidal effect at 7 mg copper sulfate per mother when administered intravenously to pregnant hamsters.

Several Group II metals are teratogenic. Beryllium has been shown to interfere *in vitro* with the differentiation of developing embryonic tissue and hence has potential teratogenicity. Hypermagnesemia develops in newborn infants if their mothers have been treated with magnesium sulfate (Lipsitz and English, 1967); unequivocal and direct involvement of magnesium in mammalian teratogenesis has not yet been established. Skeletal and feather abnormalities are observed effects of excess calcium in bird embryos (Grabowski, 1966).

Zinc deficiency causes congenital malformations in chicks (Blamberg et al., 1960) and in pregnant rats (Hurley and Swenerton, 1966). Zinc excess in the maternal diet increases the incidence of hydrocephalus in rats (O'Dell, 1969) and in hamster embryos (Ferm and Carpenter, 1968). Cadmium is embryotoxic to rodents irrespective of the mode of administration. Fetal resorption, obliteration of facial structure, and clefts in the lip and palate are some of the teratogenic manifestations in hamsters (Ferm and Carpenter, 1967). Zinc protects against cadmium-induced teratogenesis in hamsters. Cobalt decreases toxic effects of cadmium in tissues such as the testes, but it cannot prevent teratogenesis of cadmium. Cadmium at 10 ppm in the drinking water of mice was found to be teratogenic, resulting in abnormal and sterile offspring. Fetal resorption and skeletal, heart, or kidney abnormalities in stillborn or crippled offspring are manifestations of cadmium teratogenicity in rats (Scharpf et al., 1972). Fetal resorption or abnormal and short-lived offspring are the signs of cadmium teratogenicity in goats (Anke et al., 1970). The manifestations of cadmium teratogenesis were dose-related in all species.

The teratogenic effects of mercury and its compounds in humans and experimental animals have been established beyond doubt. Organic mercurials and methyl mercury salts have greater teratogenicity than have inorganic mercury salts; methyl mercury chloride, ethyl mercury acetate, mercuric sulfide, and mercuric phosphate provoke teratogenesis. The effects are far-reaching, irrespective of the mode of administration, including vaginal application. The developing nervous and hematopoietic systems of the fetus are very susceptible to teratogenesis from methyl mercury salts. Fetal resorption and death, incidence of cleft palate, reduced body weight of newborns, abnormal tails, and abnormal nervous tissues are some of the teratogenic manifestations of mercury. The abnormal changes in the nervous tissue of neonates due to mercury teratogenesis include malformation of the spinal cord, retarded growth of the cerebellum, incomplete development of cerebral granular matter, diminution and atrophy of cells in the cerebral white matter, and disappearance of cerebellar granular cells (Krehl, 1972). Some of these teratogenic effects were partially present in the unborn fetuses of women suffering from Minamata disease. Severe mental retardation, seizures, and mild spasticity were reported in the children born of Japanese mothers who were exposed to mercury intoxication; these mothers showed none of the symptoms enumerated above. Birds fed mercury chloride laid infertile eggs with reduced hatchability (Cogburn et al., 1973).

Among the metals of Group III, subgroup B metals do not cross the placental barrier and hence appear to be nonteratogenic. Aluminum, indium, and thallium of subgroup A cross the placental barrier, and indium and thallium are teratogens.

Excessive dietary aluminum salts cause growth retardation in second- and third-generation mice without any other significant teratogenic effects. Resorption of the fetus, polydactyly, and growth retardation are the main symptoms of indium teratogenesis in golden hamsters (Ferm and Carpenter, 1970). Indium appears to be a potent site-specific teratogen, since it affects enzyme-dependent limb-bud differentiation. Epidemiologic evidence (Moeschlin, 1965) suggested the involvement of thallium salts in human teratogenesis; thallium mimics potassium and accompanies potassium across the placental barrier. Chronic thallium intoxication in early pregnancy causes limb deformities in babies; the central nervous system of the baby is affected if the mother is exposed in later periods of pregnancy. There are no substantive experimental reports about thallium teratogenicity in experimental animals.

Among Subgroup IVA metals, lead and its salts exhibit teratogenicity and embryocidal effects. The cationic salts of germanium, tin, and Subgroup IVB metals do not appear to pass through the mammalian placental membrane. Although there are no reports, anionic germanates and titanates can cross the barrier. Dietary lead was found to be highly teratogenic in mice; the strain died within two generations; fetal males were more vulnerable than females (Schroeder and Mitchener, 1971). Polydactyly was observed in golden hamsters following dietary ingestion and parenteral administration of soluble lead salts (Ferm and Carpenter, 1967). Lead encephalopathy in neonatal man or animals is due mainly to the transfer of lead through maternal milk, not to transfer across the placental membrane. In birds, malformation of the anterior central nervous system was induced by lead teratogenesis.

Arsenic is the only teratogenic member of the Group V metals; it is also embryocidal. In the anionic form, both arsenic and vanadium pass through the mammalian placental membrane. Pentavalent arsenic is teratogenic in rodents.

Among Group VI metals, anionic forms of chromium, molybdenum, selenium, and tellurium can cross the mammalian placental membrane. Molybdenum, selenium, and tellurium appear to exhibit mild teratogenic and no embryocidal effects. Epidemiologic evidence involves selenium in teratogenesis in pigs, sheep, and cattle whose young are born with deformed hooves. Malformations among chickens and sheep raised in seleniferous areas have been reported (McLaughlin *et al.*, 1962). Birth of a deformed infant and miscarriages in several pregnancies among female laboratory technicians exposed to selenium suggest teratogenic potential of selenium among humans (Robertson, 1970). Experimental evidence in hamsters injected intravenously with 2 mg per kg of selenium as selenite proved to be negative for selenium teratogenicity. Selenium prevents cadmium- and arsenic-induced teratogenesis (Holmberg and Ferm, 1969).

Metallic tellurium induced hydrocephalus in fetuses and in newborn when pregnant rats were fed 3000 ppm of tellurium (Duckett, 1972). Molybdate did not exhibit any teratogenic effect in pregnant hamsters. Severe demyelination of the central nervous system was observed in newborn lambs, however, when the pregnant ewes were fed a diet high in molybdate (Mills and Fell, 1960).

Among the Group VII metals, only manganese is known to affect embryos. Manganese deficiency causes teratogenesis in rodents, especially skeletal malformations (Hurley, 1968), and malformation of beaks in chickens (Lyons and Insko, 1937). Excess manganese proves to be embryocidal and not teratogenic in pregnant hamsters.

Among Group VIII metals, soluble iron, cobalt, and nickel ions can pass through mammalian placental membranes; other metals of this group and their salts do not. None of the metals of this group is reported to have any teratogenic effect.

6

SUMMARY AND OVERVIEW OF METAL TOXICITY

Previous chapters have indicated the physicochemical and physiologic basis for metal toxicity. An understanding of the composition and dynamic aspects of our geochemical environment provides an overview of the interplay between man and his spaceship Earth. We are becoming more aware that the geologic changes in climate may affect the cycles of specific metals, as illustrated by the mercury cycle (Fig. 6-1). Similar cycles that are drastically affected by man's utilization of earth's geologic resources could be drawn for selenium, arsenic, and other metals (Wood, 1974). Concepts of geologic, industrial, agribusiness, and urban utilization and pollution must be incorporated into the overview of metal toxicity. Man's activities have increasingly disturbed geologic caches of heavy metals. Relatively high concentrations of metals, such as cadmium, mercury, indium, thallium, and lead, have been released into the environment from the earth's reservoirs of immobilized ores and minerals. Consequently, man and animals are increasingly exposed to less-common metals at an alarming rate. The contributions of mineral toxicities to disabilities or diseases of man and other mammals are often nebulous. A cause–effect relationship is one of numerous questions being considered from both experimental and epidemologic approaches (Hopps, 1972).

Volume 2 presents a review of the occurrence, exposure, chemistry, biological activity, absorption, metabolism, excretion, toxic symptoms, detoxication, and homeostatic mechanisms of metals with no significant radioactivity. That review includes most of the available reliable data on metal toxicity. Those data are incorporated into generalizations in this volume to present the overall toxicity. These generalizations obviously are subject to corrections and additions with the accumulation of more reliable

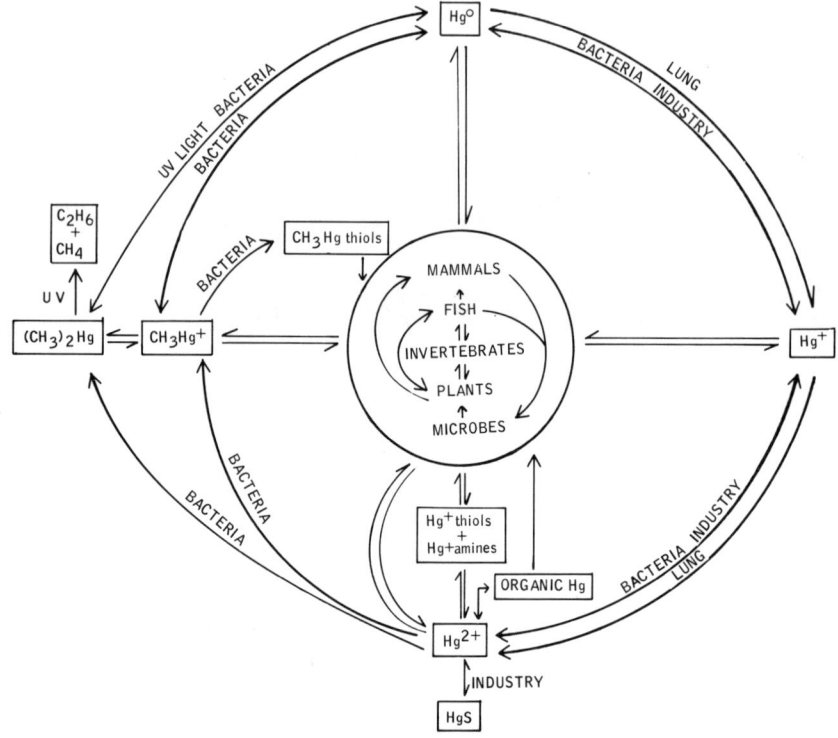

FIG. 6-1. Ecologic mercury cycle.

data. Much of the earlier data on metal toxicity suffers from poor standardization and the presence of trace impurities in the samples tested. Additional information should substantiate these generalizations, not refute them.

Specific references for the material summarized in this chapter are given in Volume 2 (Venugopal and Luckey, 1977). More details on these and other patterns of toxicity of the nonradioactive metals of the periodic table are summarized in an overview of comparative metal toxicology in the last chapter of Volume 2.

Harmful effects can be observed with excessive administration of any substance. An inert substance becomes harmful when an excess displaces essential nutrients or calories. A relatively innocuous substance may cause little reaction until a level is reached at which colligative properties will cause osmotic disturbance. Many heavy metal compounds are so insoluble that they cannot be absorbed from the digestive tract in quantities large enough to be toxic; however, parenteral injections of the same materials reveal an inherent toxicity. The characteristics of each metal compound dictate specific harmful effects, depending on dosage, mode of administration, condition of the recipient, and the time used in the evaluation of toxicity.

Toxicity is the capacity of any chemical to affect adversely the activity of living organisms. Early mortality, retarded growth, impaired reproduction, neonatal mortality, neoplasms, and varied disease symptoms are common criteria for metal toxicity in mammals. Adverse effects include behavioral changes in individual organisms and ecologic changes that affect collective populations. Excessive doses of nutritionally essential metals can cause adverse effects. The condition of the exposed subject is of great importance in the response to toxic doses of metals. Under pathologic conditions, such as renal or lung disorders, biliary obstruction, breakdown of homeostasis, or physiologic stress, metals may accumulate more readily and with more toxic effects. The response to a given dose will vary with time of day, degree of stress, or the physiologic state of the recipient.

Metals and their salts exert toxic effects in the organism at tissue, cellular, subcellular, and molecular levels. Toxicosis at the cellular level causes deranged reproduction, differentiation, and maturation, as exemplified in teratogenesis. Some metals affect the permeability of cell membranes and disturb energy metabolism; other metals decrease the stability of lysosomal membranes to disrupt cell functions by releasing various hydrolases. Colligative properties are important in the toxicity of some metals. At the molecular level some metals interact with proteins, leading to denaturation, precipitation, allosteric effects, or enzyme inhibition; other metals bind to nucleic acids, leading to irreversible conformational changes. Interaction with DNA can cause mutation or carcinogenesis. Metals cause acute, chronic, latent, and recondite toxicosis, depending on the susceptibility of the animal, and on the dose, the mode of administration, and the duration of exposure. Acute or chronic toxicosis can be produced in experimental animals and is readily detected in man, whereas symptoms associated with recondite toxicity are difficult to produce experimentally and are difficult to diagnose in man.

Delayed, or latent, toxicity, in which clinical symptoms are observable only months or years following exposure, is caused by metals such as chromium and beryllium; latent toxicity is based on epidemiologic evidence and has been proven experimentally. The lethal doses of a metal compound for acute, chronic, recondite, and latent toxicity vary, and they depend on biologic and environmental variations. The inherent toxicity of a metal depends on its capacity to disturb the dynamic equilibrium of life processes in biologic systems by combining with cell organelles, macromolecules, or metabolites.

The inherent toxicity of a metal is an expression of its electrochemical character; the solubility, stability, and reactivity of its compounds in body fluids and tissues; its ability to chelate with ligands of biologic macromolecules and the stability of these metal chelates; and, finally, its physical form—ionic, colloidal, or radiocolloidal—in susceptible target tissues. This inherent toxicity is increased by a higher rate of absorption from the alimentary or respiratory tracts, a higher rate of absorption into the blood from the sites of parenteral injections, and a higher rate of distribution from the blood to the various susceptible tissues. Inherent toxicity is attenuated by the efficiency and capacity of the organism's homeostatic mechanism that controls the absorption, distribution, retention, detoxication, and excretion of the metal ion.

The following pages give a brief summary of the toxicity of individual metals under the eight groups of the periodic table; this presentation summarizes the detailed presentation of metal toxicity in Volume 2.

SUMMARY OF METAL TOXICITY BY GROUP

Group I Metals

Metals of subgroup IA form water-soluble, polar compounds; their toxicity increases with increased atomic weight and electropositivity, cesium being the most toxic metal of the subgroup. The toxicity of rubidium and cesium results from their competition with and displacement of sodium and potassium in cell reactions. Lithium occupies a unique place in the toxicity scale in this subgroup, due to its small ionic size and high charge-to-mass ratio. The caustic actions of strong solutions of the alkali hydroxides result in acute toxicosis from tissue dissolution. In contrast, metals of subgroup IB are less electropositive than subgroup IA metals and are more toxic. Within subgroup IB, gold has potential toxicity, but the poor solubility of gold salts results in negligible absorption from the alimentary tract and renders gold less toxic than copper or silver. Gold salts tend to reduce to elemental gold in tissues. The greater solubility of copper salts in tissue fluids renders copper more toxic than silver when administered intravenously. Silver is immobilized by reaction with –SH groups of tissue proteins following parenteral administration, while copper ions are rapidly absorbed. Divalent copper is more toxic than monovalent copper.

Subgroup IA Metals

Lithium. Cationic lithium may be the most toxic of alkali metal ions when taken orally; the toxicity of other ions varies, depending on the anion and other factors. In mammals the gastrointestinal, renal, neuromuscular, central nervous, cardiovascular, and endocrine systems are affected by lithium toxicity, which leads to disability and death. Osmotic disturbance in the digestive tract causes gastroenteritis, anorexia, and nausea. Derangement of neuromuscular activity results in tremor, ataxia, and hyperactive reflexes. Symptoms of lithium toxicity in the central nervous system are slurred speech, blurred vision, dizziness, convulsions, and stupor. Polyuria, glycosuria, and weight gain due to water and salt retention are due to renal malfunctioning. Lowering of the potassium level in heart muscle causes pulse irregularities, circulatory failure, and collapse. Goiter and decrease in thyroid function are the major lithium toxicity symptoms in the endocrine system. Lithium causes teratogenicity in experimental animals; the symptoms include skeletal abnormalities, high incidence of resorption, and decreased fetal survival and litter size. The mechanism of lithium toxicity is mainly attributed to inhibition of adenyl cyclase and cAMP formation, derangement of catecholamine metabolism and cerebral carbohydrate metabolism, and changes in the cellular microenvironment. High salt intake alleviates lithium toxicity symptoms. There are no specific detoxication or homeostatic mechanisms for lithium, although homeostatic mechanisms for sodium or potassium can handle some lithium excesses.

Sodium. The major symptoms of acute sodium toxicity are nephrosis, hypertension, anemia, lipemia, severe proteinemia, inhibition of spermatogenesis, epileptiform convulsions, somnolence, and stupor. Sodium salts are not generally toxic, due to an efficient homeostasis that operates at the renal excretory level, and is controlled by aldosterone and posterior hypophyseal antidiuretic hormones.

Potassium. Symptoms of acute oral potassium toxicity in mammals include tonoclonic convulsions, diarrhea, acute gastroenteritis, cardiac and central nervous system depression, flaccid paralysis of the extremities, and necrosis of the renal epithelium. Potassium salts are toxic only at distinctly unphysiologic doses. Radiation from ^{40}K, 20 mrems per year in adults, may be oncogenic in the aged.

Rubidium. Parenteral administration of rubidium salts is dangerous. Acute symptoms of rubidium toxicity are violent tetanic spasms and convulsions. Growth retardation, impaired reproduction, and decreased longevity are the symptoms of chronic intoxication. Rubidium has depolarizing effects on cell membranes and causes profound changes in intracellular cation composition. These effects cause acidosis, neuromuscular effects, and myocardial irritability.

Cesium. Acute cesium toxicity causes liver disorder, neuroendocrine disturbances, and neuromuscular toxicosis leading to irritability and convulsions. The mechanism of cesium toxicity appears to be similar to that of rubidium toxicity. Cesium and rubidium can substitute for potassium to a level of 10% in the diet, beyond which the heavier alkali metal ions disrupt cellular function and cause toxic symptoms.

Subgroup IB Metals

Copper, silver, and gold, metals of subgroup IB, form coordination complexes since the empty d-orbitals in their atoms are available for hybridization; the metals have great affinity for sulfur- and nitrogen-containing ligands. Copper is present in a number of enzymes involved in oxidation and reduction. Silver combines with proteins to form stable insoluble complexes that finally decompose into black silver sulfide; silver salts tend to decompose in the tissues. Metals of subgroup IB are absorbed to a limited extent from the gastrointestinal tract, and their excretion is mainly fecal. Copper transport to and from tissues is well organized, while silver and gold transport is poor.

Copper. The range between nutritive requirement and toxicity of copper is wide. Symptoms of chronic copper toxicity in mammals are sporadic fever, vomiting, epigastric pain, diarrhea, and jaundice. Acute copper toxicity causes hypotension, hemolytic anemia with intravascular hemolysis, uremia, and cardiovascular collapse. Inhalation of dusts of copper salts causes ulceration and perforation of the nasal septum and pharyngeal congestion. Copper toxicity also causes necrotic hepatitis and brain tissue damage; copper toxicity is greater with low dietary intake of molybdate, sulfate, zinc, and iron. Wilson's disease, a disorder of copper metabolism in humans, is inherited as an autosomal recessive trait; copper accumulates excessively in the liver, cornea, and brain and causes hepatic cirrhosis, necrosis, and sclerosis of the corpus striatum and trauma in the brain. Copper ions increase lysosomal fragility to release acid hydrolases and cause cell degeneration. Copper toxicity increases cellular permeability in erythrocytes, leading to lysis, or causes agglutination of erythrocytes; inhibition of glutathione reductase in erythrocytes and loss of intracellular reduced glutathione occur, as do mitochondrial swelling and disturbance in energy metabolism. Efficient copper homeostatis occurs in the liver and in the intestinal mucosa. Storage of acid drinks (i.e., whisky sour) in metal containers results in 1–2 outbreaks of acute poisoning each year in the United States (C.D.C., 1975).

Silver. Colloidal silver is quite toxic following parenteral administration; it causes tumors in the liver and spleen. Silver salts cause corrosion of intestinal and gastric mucosa. Argyria, the black pigmentation of the tissue, is the deposition of the silver–protein complex or silver sulfide as opaque granules in the elastic fibers of the tissue. Acute dietary silver toxicity causes severe gastroenteritis, diarrhea, hypotension, spasms, paralysis, and respiratory failure. Chronic silver toxicity symptoms are fatty degeneration of the liver and kidney, and liver necrosis. Silver ions are believed to reduce the availability of selenium to nonheme iron proteins and to impair the ability of these proteins to function as electron carriers. Silver accumulates in humans with age.

Gold. Gold metal dust and its insoluble salts are not toxic to mammals via oral administration; soluble salts are toxic. Dietary absorption of gold salts is poor, and the excretion of parenterally administered gold salts is fecal and urinary in a 1:4 ratio. Toxicity symptoms are similar to those of silver and include blood dyscrasias, leukopenia, agranulocytosis, aplastic anemia, and allergic contact dermatitis. The mechanism of gold ion toxicity is its affinity for thiol groups and inhibition of thiol enzymes. The significance of gold binding to DNA in gold toxicity is not clear.

Group II Metals

Metals of Subgroup IIB are more toxic than are metals of Subgroup IIA. Within subgroup IIA, the lightest metal, beryllium, is the most toxic, irrespective of the nature of its salts or the mode of administration. The great inherent toxicity of beryllium is due to its small ionic size and high charge to mass ratio, allowing ready penetration into tissues and cells. Lack of homeostatic mechanisms for beryllium in mammals enhances this inherent toxicity. Magnesium, an essential metal for mammals, also possesses high inherent toxicity and acts by disturbing the osmotic equilibrium in biologic tissues and fluids. This inherent toxicity, however, is manifested only from parenteral administration. Magnesium salts are practically nontoxic by oral administration, since effective homeostasis operates in intestinal absorption. The inherent toxicity patterns of calcium, strontium, and barium are similar to those of magnesium; the toxicity increases with increase in electropositivity and atomic weight. Poor solubility of a heavy metal salt, such as barium sulfate, renders it nontoxic, while the soluble barium chloride is highly toxic. Metals of subgroup IIB exhibit the same toxicity pattern. The high lipid and aqueous solubility of mercury and its inorganic and organic salts enhances its great inherent toxicity.

Subgroup IIA Metals

Beryllium. Beryllium accumulation causes cellular death in all tissues, and it may cause symptoms months or years following exposure. Symptoms of beryllium toxicity are numerous, since beryllium affects many processes. Inhalation of beryllium aerosols causes a granulomatous, crippling, and incurable disease. Symptoms of chronic beryllium toxicity include pneumonitis, heart enlargement and congestive cardiac failure, cyanosis, malignant tumors, and calcified inclusions in cells and tissues. The carcinogenicity of beryllium in experimental animals is well established. The mechanism of beryllium toxicity is complex. Beryllium inhibits a variety of enzymes, including those involved in DNA synthesis. Beryllium displays a high and selective affinity for cell nuclei, and it interferes *in vitro* with the differentiation of developing embryonic tissue. There is no detoxication or homeostatic mechanism for beryllium. It is most toxic!

Magnesium. Acute magnesium toxicity causes nausea, malaise, general depression, and paralysis of the respiratory, cardiovascular, and central nervous systems. The toxic effects of magnesium on the central nervous system are due to a decreased liberation of acetylcholine at the neuromuscular junction and sympathetic ganglia. Hypotension, cutaneous vasodilation, and cardiac arrest are the major magnesium toxicity symptoms of the cardiovascular system. Depression of cardiac muscle activity and hypoxia, secondary to paralysis of respiratory tissues, are responsible for the local intraspinal and general anesthetic effects of magnesium toxicity.

Calcium. Calcium salts are not generally toxic via oral administration; calcium toxicity from intravenous administration is attributed to sudden osmotic changes. Calcium ions in excess are known to depress muscle function by increasing the threshold of excitability of muscle and nervous tissues; a state of stupor results.

Strontium and Barium. Strontium salts are less toxic than the corresponding calcium salts; acute strontium toxicity causes respiratory failure in rats, while calcium toxicity produces cardiac failure. Excessive salivation, vomiting, colic, and diarrhea are the major symptoms of acute strontium toxicity.

To the extent of their solubility, barium salts are more toxic than either calcium or strontium salts. In addition to the symptoms associated with calcium and strontium toxicity, tremor, muscular paralysis, hypertension, and paralysis of the central nervous system are the major symptoms of barium toxicity. Hypertension causes hemorrhage in the kidney and the digestive tract, and low serum potassium levels and leucocytosis result from acute barium toxicity. The strong direct stimulation of arterial muscles from excessive doses of barium salts causes strong vasoconstriction, leading to muscular paralysis.

Homeostatic mechanisms maintain normal levels of magnesium, calcium, and, to a certain extent, strontium and barium. The similarity of barium to other subgroup IIA metals in atomic structure and metabolism suggests that soluble salts of radium would exceed those of barium in their chemical toxicity.

Subgroup IIB Metals

Subgroup IIB metals are posttransition metals with a large number of electrons in the inner orbitals; they form strong covalent bonds leading to stable coordination and chelation complexes, especially with less electronegative elements. Zinc and mercury, and, to a certain extent, cadmium, form soluble salts; all three have high affinity for thiol groups: mercury > cadium > zinc. The electropositivity and toxicity of zinc, cadmium, and mercury increase with increasing atomic weights; of the three metals, cadmium is most poorly absorbed from the digestive tract. Zinc is the least toxic of the subgroup IIB metals. The toxicity of mercury is higher than that of cadmium due to its greater electropositivity, solubility, and increased absorption, and penetration into the tissues.

Zinc. Zinc salts are relatively nontoxic due to efficient zinc homeostasis. The margin of safety between the normal zinc intake and the toxic level of zinc is great. Large doses of zinc chloride are corrosive to the skin and cause acute damage to buccal and gasteroenteric mucous membranes. Growth retardation, faulty reproduction, anemia, and pancreatic fibrosis are some symptoms of chronic zinc toxicity. Acute zinc toxicity causes lassitude, bloody enteritis, diarrhea, and depression of the central nervous system, leading to tremors and paralysis of the extremities. Inhalation of zinc compounds causes metal fume fever and pneumonitis. Zinc teratogenicity is reported in hamsters. Intratesticular injection of zinc causes sterility and tumors in the testes of rats. Excess dietary zinc disturbs mineral metabolism in bone by inhibiting the absorption of copper and phosphorus.

Cadmium. Cadmium is toxic to all tissues. Symptoms of chronic cadmium toxicity include growth retardation, impaired kidney function, poor reproductive capacity, hypertension, tumor formation, hepatic dysfunction, poor lactation, and lowered hematocrit levels. Teratogenic effects are noticed in offspring when pregnant animals are exposed to chronic toxic doses of cadmium. Inhalation of cadmium salts leads to pulmonary emphysema and bronchitis. Symptoms of acute cadmium toxicity are excessive salivation, abdominal pains, diarrhea, vertigo, and loss of consciousness. Carcinogenicity of cadmium in animals has been well established by parenteral administration, although cadmium is not carcinogenic by the oral route. Along with other metals—high dietary copper, lead, and zinc—cadmium has been identified as the etiologic factor in "itai-itai-byo," which is characterized by osteomalacia. Cadmium toxicity is related to its antimetabolite activity against essential metals, copper, iron, and especially zinc, to its inhibitory effects on enzymes such as alkaline phosphatase and ATPases, and to its blocking of the renal synthesis of 1,25-dihydroxycholecalciferol. Changes in membrane permeability leading to abnormal transport of metabolites, derangement of cellular energy metabolism, and binding to cellular respiratory components are the suggested mechanisms of cadmium toxicity in mammals. Cadmium accumulates with age. The formation of inactive and nontoxic cadmium-binding proteins, such as metallothionein, following cadmium intoxication, serves as a mild detoxication mechanism for cadmium.

Mercury. Mercury and its compounds are highly toxic due to their solubility in both polar and nonpolar solvents. The volatility of the metal and its absorption through the skin render mercury an industrial and environmental hazard. The ability of microorganisms and plankton to trap mercury from industrial waste water and the accumulation of mercury in the tissues of fish result in efficient recycling of mercury from the environment to redistribution into food products. Since man's activities have increased the level of mercury in the environment, mercury intoxication has become a health hazard. Organic mercury salts, such as monoalkylmercuric chloride, are more toxic than other organic mercury compounds or inorganic mercury salts. The enterohepatic absorption and lipid solubility of methyl mercury accounts for its preferential retention in tissues and its relatively long biologic half-life. Although mercury compounds have been reported to be stimulatory, they are not required for any specific biologic function in living tissues.

The absorption of mercury compounds is very rapid from all modes of intake. The absorbed mercury ions penetrate most tissues rapidly to exert their toxic effects on target tissues. An indirect, possible detoxication mechanism for mercury is the enhanced formation of liver metallothionein, which can bind mercury ions.

Mercury and its inorganic salts are protoplasmic poisons lethal to all species. Symptoms of acute mercury intoxication include nausea, headache, abdominal pain, diarrhea, metallic taste, and albuminuria. Stomatitis and gingivitis develop with the swelling of salivary glands and ulceration in the buccal cavity. Hemorrhagic colitis, hemolysis, digital tremors, delirium, and hallucinations precede death, which is caused by extreme exhaustion. Symptoms of chronic mercury intoxication are less severe and slow in appearing, but equally devastating. Ataxia, dysarthria, dysphagia, ocular lesions, uncoordinated movements of arms and legs, impaired hearing, and impairment of taste and smell are early symptoms of chronic mercury intoxication. Organic alkyl mercury salts affect the central nervous system and are teratogenic. At the molecular level, the effect of mercury intoxication results from strong inhibition of sulfhydryl enzymes, which subsequently affects cellular metabolism in all tissues. Mercury intoxication studies with isolated cells show changes in cell membrane permeability, loss of cellular potassium, reduced uptake of glucose, decreased electrical potential across the cell membrane, and a strong inhibition of cellular respiratory enzymes.

The far-reaching teratogenic effects of organic mercuric salts include fetal resorption and death, lowered body weight of offspring, mental retardation, incidence of cleft palate, and changes in the nervous tissue.

Selenites and tocopherol provide some protection against mercury intoxication by causing a shift in tissue mercury distribution.

Group III Metals

Metals of Group III possess the highest inherent toxicity among the metals, but this toxicity is attenuated by the behavior of the soluble salts of these metals at physiologic pH. Soluble salts are hydrolyzed to the corresponding hydroxides; these hydroxides are insoluble in tissue and fluids and separate as stable precipitates or colloidal or radiocolloidal particles. Thallium is the only exception to this behavior; thallium exists as a stable monovalent ion that behaves metabolically like K^+ in its absorption from the alimentary tract and subsequent distribution to tissues. Thallium and selenium are the most toxic of all nonradioactive metal ions. The inherent toxicity of the other metals in this group can be demonstrated by preparing their stable hydrated ionic forms and administering these ions parenterally into experimental animals. The hydrolysis of the soluble salts of the metals of Group III at the tissue pH levels is more pronounced in subgroup B (lanthanides) than in subgroup A. The lanthanide hydroxides are not soluble at the pH of the intestinal fluids and are scarcely absorbed. They are used as unabsorbed nutritional markers. The oxides and hydroxides of lanthanides behave as nontoxic or slightly toxic compounds when administered orally. Soluble salts of lanthanides are toxic when given intravenously, however, because the hydrolyzed products, the hydroxides, accumulate in the reticuloendothelial system and block its normal functioning. Gallium and indium salts are effective toxicants by this route. Aluminum and the lanthanides are considered to be practically nontoxic by oral administration; gallium, indium, and thallium are toxic. The chemical toxicity of the most highly radiotoxic metals, the actinides, is not yet fully assessed. Following parenteral administration, metals of Group III display toxicity ranging from low to very high. Monovalent thallium has the greatest chemical toxicity on a molecular basis. Metals of Group III possess inherent toxicity, which increases with increased atomic weight and electropositivity. Metals of subgroup IIIA (aluminum excepted) exhibit greater toxicity than do metals of subgroup B.

Subgroup IIIA Metals

Aluminum. Skin lesions, nervous afflictions, gastrointestinal disturbance, growth retardation, perihepatic granulomas, and fibrous peritonitis are some common symptoms of aluminum toxicity. Phosphate depletion in the tissues, negative phosphorus balance, lowered phosphate absorption from the digestive tract, and adverse alterations in phosphorylation reactions in the tissues are responsible for aluminum toxicity. Inhalation of finely divided aluminum dust by humans causes encephalopathy and pulmonary fibrosis; binding of aluminum with nervous tissues causes neurofibrillar degeneration.

Gallium. Cationic gallium salts are relatively nontoxic by the oral route. Intravenous injection of gallium chloride shows its high toxicity; it is sometimes immediately lethal due to formation of colloidal $Ga(OH)_3$, which leads to colloidoclastic shock. Symptoms of acute gallium toxicity in mammals include vomiting, nausea, diarrhea, anorexia, weight loss, photophobia with occasional blindness, and dermatitis. Reduced hemoglobin levels, swollen lymph nodes, nuclear fragmentation with necrosis of lymphoid tissues, and extensive renal damage are the manifestations of gallium intoxication. In addition to its normal distribution to the skeleton, kidney, and liver, gallium, especially in the form of gallium citrate, accumulates in a variety of solid tumors following parenteral injection; thus, ^{67}Ga citrate or similar organic complexes are useful in locating malignant tumors. Parenterally administered $Ga(NO_3)_3$ is reported to inhibit growth of solid tumors in man and animals.

Indium. Cationic indium salts exhibit low toxicity by the oral route due to their poor absorption from the digestive tract. Soluble indium salts are highly toxic following parenteral administration. An ionic form of indium, suitably buffered to prevent hydrolysis to insoluble $In(OH)_3$ at physiologic pH is less toxic than an aqueous solution of $InCl_3$. Specially prepared colloidal In_2O_3 is about 250 times more toxic than is regular In_2O_3. In addition to being distributed in the skeletal and reticuloendothelial systems, indium salts accumulate in neoplasms.

Symptoms of acute indium intoxication are anorexia, localized convulsive motions, hind-leg paralysis, pulmonary edema, necrotizing pneumonia, and renal and hepatic damage with resultant dysfunction. Chronic indium intoxication leads to weight loss, poor growth, and extensive necrotic damage to the liver and kidneys. The reticuloendothelial system is very drastically affected. Indium salts are direct calcifiers and cause topical calcification. Indium is a potent site-specific teratogen in hamsters, and it affects enzyme-dependent limb-bud differentiation. Indium is also embryopathic.

Thallium. Thallium, which is situated between mercury and lead in the periodic table, is the most highly toxic cumulative cation (excluding radionuclides). Thallium is the most basic among subgroup IIIA metals; its salts are soluble and are not hydrolyzed to oxides at physiologic pH. Thallium resembles and behaves like potassium in its metabolism, accumulation in erythrocytes, active transport, excretion, and soft-tissue distribution. Thallium has undefined affinity toward −SH groups and sulfur-containing ligands, but does not generally block −SH groups in some thiol-containing enzymes and proteins.

Thallium salts are easily absorbed from the digestive tract. Thallium excretion is slow and prolonged; it is mainly urinary and partly fecal, with lesser amounts in sweat, milk, and tears. Thallium intoxication causes polyneuritis, digestive system disorders, encephalopathy, alopecia, and retrobulbar neuritis; symptoms of acute thallium intoxication vary with age, mode of administration, and animal species. The common symptoms are gastrointestinal hemorrhage, gastroenteritis, metallic taste, and nausea. Neurologic symptoms include delirium, hallucinations, convulsions, tingling pain in the extremities, muscular weakness, coma, and puffiness of cheeks, eyelids, and lips. Symptoms of chronic thallium intoxication are mild and include, in addition to the acute symptoms named above, alopecia, cardiac disorders such as hypertension, irregular pulse, and renal disorders. Thallium acts as a mitotic agent and general cellular poison and is teratogenic.

Subgroup IIIB Metals

Scandium. Typical of the nonradioactive subgroup IIIB metals, scandium salts are practically nontoxic and are used as nutritional markers. Administration of lethal doses of scandium salts to experimental mice causes abdominal distention, depressed respiration, and sedation, followed by death. Intravenous Sc^{3+} causes respiratory paralysis and cardiovascular collapse. Chronic scandium intoxication is reported to cause growth retardation and may be tumorigenic.

Yttrium. Yttrium toxicity symptoms include anorexia, asthenia, and depression of physiologic functions, with death due to cardiac and respiratory failure. Citrate complexes of yttrium are more toxic than are simple salts. Subcutaneous implantation of metallic yttrium induces growth of granulomatous tissue.

Lanthanides. The lanthanides are a subgroup of 15 inner-transition metals with similar physical and chemical properties and with very small differences in their atomic weights; they occupy one space in Group IIIB of the periodic table. These metals exhibit the general properties of Group III, including hydrolysis of water-soluble salts at physiologic pH into insoluble hydroxides or oxides. These hydroxides separate as fine suspensions, colloidal, or radiocolloidal forms, depending on the concentration of these salts. These metals are not essential to mammalian nutrition: some are stimulatory. They are relatively nontoxic when compared with other metals.

Lanthanide salts are poorly absorbed from the digestive tract due to insoluble phosphate or hydroxide formation; however, within physiologic limits lanthanide citrate and EDTA complexes are absorbed rapidly and completely. Lanthanide salts are distributed rapidly into the liver, kidney, and reticuloendothelial system, with gradual uptake and retention in the bones. The excretion of these metal compounds is both fecal and urinary and depends on the dose and mode of uptake. Simple inorganic lanthanide salts are less toxic than are the citrate or EDTA complexes. The light lanthanides are slightly more toxic than the heavy lanthanides; the intermediate lanthanides are the least toxic of the three subdivisions. A biphasic mortality response is exhibited by these salts following intraperitoneal administration. Symptoms of acute lanthanide intoxication are immediate defecation, writhing, ataxia, sedation, and labored respiration. Respiratory and cardiac failures cause death. Pulmonary edema, hyperemia, liver edema and necrosis, pleural effusion, and granulomatous peritonitis with serous and hemorrhagic ascites are the clinical symptoms. Acute chemical pneumonitis, subacute bronchitis, and focal hypertrophic emphysema occur following the inhalation of lanthanide fluorides and aerosols. Subcutaneous implantation of yttrium, cerium, praseodymium, gadolinium, dysprosium, or ytterbium metal pellets causes granulomatous tissue which may become tumors. Lanthanide salts cause soft-tissue calcification at the sites of injection; systemic calficication in the spleen develops following intravenous injection.

The light lanthanides are very effective anticoagulants and also cause greater hepatic damage than do the heavy or medium lanthanides. Lanthanum is the most toxic of the lanthanides because of its high electropositivity, high charge density, and tendency to form complexes through electrostatic bonding. Hepatic function is affected due to accumulation of neutral lipids, mitochondrial damage, and perinuclear vacuolization; the detoxicating capacity of the liver is reduced. The catecholamine content of brain and adrenal tissues increases. Instillation into the eye causes corneal ulceration. Inhalation of europium salt aerosols increases the ratio of large to small lymphocytes; europium chloride produces teratogenic effects in chickens.

Actinides. The actinides constitute a series of 15 highly radioactive metals and are very similar to the lanthanides in electronic configuration and in general chemical properties. The actinides are bone seekers and behave like the lanthanides in absorption from the digestive tract, tissue distribution, and slow secretion. The toxicity associated with the chemical nature of these metals is not fully assessed due to their intense radioactivity; their deposition and transient retention in the bone enhances their high radiotoxic property, since they continue to exert their intense radiation damage on the skeleton. Poor gastrointestinal absorption, formation of particulate or colloidal forms following absorption or administration by other routes, and subsequent slow removal through phagocytosis by the reticuloendothelial system may account for the low chemical toxicity of the actinides. They are potent radiocarcinogens.

Thorium salts are toxic to experimental animals by causing chemical peritonitis, abdominal distention and rigidity, and intense radiation damage.

Uranium salts cause toxic effects chemically and radiologically. Uranium accumulates in the bone. Growth retardation and renal and hepatic malfunction are the major symptoms of chronic uranium intoxication. Destruction of renal tissue following acute uranium intoxication leads to death. Uranyl salts are more toxic than are tetravalent uranium compounds. The chemical toxicity of uranium compounds is attributed to changes in cellular membrane permeability by the binding of uranyl ion to phosphate ligands and to the inhibition of glucose transfer.

Transuranium Metals. The transuranium metals are radiocarcinogens. Their affinity toward phosphate ligands renders them avid bone seekers, especially the endosteal surface, from where they irradiate the neighboring cells, which become the sites of cancer induction. Osteogenic sarcoma and bone necrosis are caused by this irradiation.

The absorption of salts of transuranium metals from the digestive tract is poor; absorption from parenteral sites is also poor. Following inhalation, the metal complexes are engulfed by alveolar epithelium and sequestered in the foci of fibroses. This type of retention in the lung, in the sites of parenteral injection, and in the skeleton provides the "hot spots" of irradiation. The slow excretion of absorbed transuranic metal complexes is both fecal and urinary.

Group IV Metals

Metals of Group IV do not fully exhibit their inherent toxicity due to hydrolysis of their soluble salts at tissue pH levels and subsequent olation. The solubility of the cationic salts of Group IV metals is generally poor. Wherever the solubility is high, the metal salt exhibits its inherent toxicity (e.g., lead acetate or carbonate). Group IV metals exhibit electroneutral and amphoteric character with variable valency. Tetravalency is more common with the anions of the lighter metals such as germanium and titanium. The higher the valency, the lower is the toxicity of these two metals due to the rapid excretion of salts with higher valency. Within subgroup IVA, the comparative electropositivity, stability, solubility, and toxicity of simple cationic salts increase as the atomic weight increases. Soluble Sn^{4+} salts possess more inherent toxicity than do lead salts; the diffusibility of tin salts across biologic membranes is lower than that of lead salts; the number of soluble stannic salts is very limited, whereas soluble Pb^{2+} salts are numerous. The toxicity of the Sn^{4+} salts depends on the anionic part of the salt. It is very difficult to evaluate the inherent toxicity of subgroup IVB metals. In aerosols they can cause pulmonary granulomata. The salts of these metals exhibit low toxicity; they undergo olation and readily form insoluble and hydrated polymeric basic salts. Among Group IV metals, lead appears to be highly toxic due to the solubility of its salts and their easy diffusibility across biologic membranes; toxicity data on inorganic lead compounds are far more numerous than are data on the salts of any other metal in this group.

Subgroup IVA Metals

Germanium. Germanium hydride is the most toxic of inhaled germanium compounds. Chronic germanium toxicity symptoms in experimental animals are mainly fatty degeneration of the liver and inhibition of growth. The cathartic action of germanium salts disturbs water metabolism, leading to dehydration, depressed blood pressure, and hypothermia.

Tin. Tin and its inorganic salts have low oral toxicity due to their poor gastrointestinal absorption; cationic tin salts are soluble in acid pH, while anionic tin salts undergo hydration at physiologic pH to insoluble complex stannic acids. Intravenous injection of cationic tin salts, if they are not complexed with citrate, will produce colloidoclastic shock. Inhaled tin salts are absorbed very slowly from the lung. Parenteral administration of cationic tin salts to experimental animals causes spasms and fatal paralysis, with bleeding in the liver and central nervous system. Excess soluble cationic tin salts in the diet cause severe growth retardation due to poor food intake, decreased food efficiency, and anemia. The last could be prevented by high dietary iron and copper. Chronic tin toxicity produces testicular degeneration, severe pancreatic atrophy, and vacuolar changes in renal tubules. Divalent tin salts are more toxic than are Sn^{4+} salts due to the easy binding of Sn^{2+} to –SH compounds. An efficient homeostatic mechanism for tin is believed to be operative in man in preventing gastrointestinal absorption of tin. Nevertheless, accidents from canned foods makes tin of particular interest. Knorr (1975) reviewed tin toxicity from foods, medicants, and dermal contact.

Lead. The toxicity of lead and its compounds is well known and extensively documented and reviewed. Exposure of man and animals to lead toxicity from the environment, food, water, and inhalation of cigarette smoke is increasing. Lead is a cumulative poison that causes both chronic and acute intoxication. Chronic exposure to lead results in its deposition and immobilization in the bone, from where lead can be mobilized following metabolic disturbance.

The gastrointestinal absorption of lead is increased by low dietary calcium and high dietary vitamin D. The absorption of inhaled lead salts is rapid and complete. Lead permeates the placental barrier; lead toxicity is related to the levels of diffusible lead and the lead content of soft tissues, not to the total body content of lead. The partitioning of lead among target organs and the levels of diffusible lead in soft tissues are directly responsible for the toxic symptoms of lead. Low levels of iron and calcium increase lead toxicity by increasing the concentration of diffusible and mobile lead in the soft tissues.

Acute lead toxicity symptoms in man are lassitude, vomiting, loss of appetite, uncoordinated body movements, convulsions, stupor, and coma. Chronic lead toxicity symptoms are lassitude and vomiting, but the real damage is subtle, and the symptoms take a long time to appear. They are renal malfunction, anemia, brain and liver damage, cancer, hyperactivity, and general psychologic impairment. Children suffer permanent damage to the central nervous system. Toxic effects of lead in experimental animals include reduced growth and longevity, impaired renal and reproductive functions, splenomegaly, damage to hemopoietic and central and peripheral nervous systems, premature loss of teeth, and reduced immune capacity. Subtoxic levels of lead over a long period of time cause subtle changes in the permeability of cells of different organ tissues. Inhibition of spermatogenesis, irregularity of estrus, ovarian follicular cysts, and reduction of corpus lutea are caused by subtoxic doses of lead. Chronic and steady low-level exposure to lead causes nephritis, whereas acute intense exposure causes functional kidney breakdown. Lead encephalopathy is a disease of neonatal man or animals caused by the transfer of lead through maternal milk; the cerebellar and cerebral cortex are damaged, and brain edema with serious exudation and derangement of cellular energy metabolism occurs following changes in the permeability of the capillary cell walls. The toxic effects of lead on hemopoiesis are abnormal and fragile circulating erythrocytes and impairment of hemoglobin formation. In humans, toxic doses of lead interfere with the phagocytic activity of polymorphonuclear leukocytes and reduce immune resistance.

Lead binds $-SH-$, $-PO_4-$, and $-COOH$-containing ligands of macromolecules and biologic membranes. Increasing levels of dietary protein, ascorbic acid, nicotinic acid, calcium, iron, and phosphates decrease the susceptibility of mammals to lead intoxication.

Subgroup IVB Metals

Titanium. Cationic titanium and soluble titanates are relatively nontoxic. Cationic titanates are poorly absorbed from the mammalian alimentary tract. Soluble titanates cause slight impairment of reproductive function in experimental rats. Certain titanates are carcinogenic.

Zirconium. Zirconium salts are relatively nontoxic via dietary sources due to poor absorption. Intravenously injected cationic salts form insoluble colloidal polymers and are phagocytized. Bone-seeking zirconium is deposited and immobilized in the skeleton. Parenterally injected zirconates are more toxic than dietary zirconates.

Hafnium. Hafnium salts are also considered to be nontoxic or mildly toxic. Excessive intake of hafnium causes hepatic damage and respiratory paralysis.

Group V Metals

Salts of Group V metals are more toxic when administered parenterally than when ingested. Metals of this group exhibit varied levels of toxicity, depending on the nature of the salt. As a group, these metals should exhibit greater toxicity with increases in atomic weight and electropositivity. The poor solubility of the cationic salts and their hydrolysis to insoluble oxysalts or oxycationic salts prevent the heavier metals from exhibiting their inherent high toxicity. The lighter metals (arsenic and vanadium) in their soluble anionic form and the anionic niobates are more toxic than are the heavier metals; these anionic salts act as antimetabolites for phosphate in metabolic reactions. The metals of Group V react with $-SH$ groups of active molecules and prevent their normal functioning.

Inhalation of hydrides of these metals is lethal due to hemolysis of erythrocytes.

In general, the anionic forms of the metals of subgroup VA are more toxic than are the cationic salts, due to their greater solubility. Organic antimony compounds are more toxic than are similar lead compounds. Metals of subgroup VB are less toxic than their counterparts in subgroup VA. Among the metals of subgroup VA, trivalent compounds are more toxic than are pentavalent compounds; the reverse is true for the metals of subgroup VB.

Arsenic, the lightest metalloid of subgroup VA, is the most toxic. Trivalent anionic arsenic in the form of soluble arsenites and As_2O_3 exhibits great toxicity.

Subgroup VA Metals

Arsenic. Symptoms of acute arsenic intoxication are nausea, diarrhea, severe abdominal pains, skin eruption and inflammation, and acute and severe irritation of nose, throat, and conjunctiva. Chronic intoxication causes retardation of growth, general weakness, prostration, muscular aching, perforation of nasal septum, ulceration in the gastrointestinal tract, peripheral neuritis, tremors, chronic hepatitis, and liver cirrhosis. Trivalent arsenic inhibits enzymes such as D-amino acid oxidases and choline oxidase. Pentavalent arsenic inhibits enzyme systems such as cytochrome oxidase and glycerol phosphate dehydrogenase; it inhibits ATP synthesis by uncoupling oxidative phosphorylation. Fetal resorption and a variety of malformations in surviving fetuses establish the teratogenicity of As^{5+}. The carcinogenecity of arsenic is not unequivocally established by experimental methods; however, increasing epidemiologic evidence definitely implicates the carcinogenic activity of arsenic.

Antimony. Soluble antimony salts are considered to be more harmful than are similar lead or arsenic salts, but soluble antimony salts are few in number. Violent vomiting due to mucosal irritation is the main symptom of acute oral antimony intoxication. The vomit contains sloughed mucosal cells and most of the toxic antimony; diarrhea and lowered respiratory rate lead to death. Myocardial edema, hyperemia, and capillary engorgement also contribute to the fatality. Symptoms of chronic antimony intoxication are dyspnea, weight and hair loss, papular eruptions on the skin, jaundice, albuminuria, damage to the heart and liver, hyperplasia of the spleen, glomerular nephritis, abnormal increase in erythrocytes, and decrease in leukocytes. Chronic inhalation of subtoxic doses of antimony salts causes interstitial pneumonitis, intraalveolar lipoid deposits, and liver and cardiac damage.

Bismuth. Ingested bismuth salts are considered to be nontoxic due to poor absorption from the alimentary tract; soluble bismuth salts are hydrolyzed to insoluble oxysalts due to olation. Following intravenous injection, soluble salts are hydrolyzed to form uncharged colloids in the blood; these colloids break up the physical equilibrium of the colloids of the body, leading to colloidoclastic shock and death. Symptoms of chronic bismuth intoxication include anorexia, gastric pains, severe headache, peripheral neuritis, and bluish-gray bismuth deposition in oral epithelial cells. Hepatitis and nephrotoxicity precede death.

Subgroup VB Metals

Vanadium. Vanadium has many characteristics associated with an essential trace metal that are yet to be fully confirmed in mammals. The difference between essential and toxic doses is small, and vanadium is classified as highly toxic among the nutritionally needed trace metals. Vanadium pentoxide gives rise to a number of oxyacid anions; these vanadates are rapidly absorbed from the digestive tract and sites of parenteral administration. Following inhalation, the absorption of vanadates from the lung into the blood is more rapid and complete than from intestinal absorption. The excretion of vanadium is both fecal and urinary. Vanadium is distributed to soft tissues, and 50% of the absorbed vanadium can be retained in the bone.

Symptoms of acute vanadium intoxication include somnolence, nervous disturbances, hindleg paralysis, respiratory difficulties, hemorrhagic enteritis, convulsions, and finally death due to cardiac and respiratory failure. The neurotoxicity affects the vasomotor center and intercardiac nervous mechanisms. Fatty degeneration of the liver, parenchymatous degeneration of the kidneys, congestion of the spleen, lungs, brain, and spinal cord, and hemorrhage of the adrenals are noted during postmortem examination. Inhalation of toxic doses of vanadium causes lacrimation, nasal bleeding, acute bronchitis, diarrhea, and respiratory and cardiac failure. Pentavalent vanadium compounds are more toxic than tetra- or trivalent compounds.

Vanadium toxicity is attributed to its ability to inhibit enzyme systems such as monoamine oxidase, ATPase, tyrosinase, choline esterase, and cholesterol synthetase. Vanadates decrease the activities of coenzymes A, NAD, and Q. Vanadium salts decrease immune resistance and allergic reactivity. Epidemiologic and statistical evidence suggests the involvement of vanadium in lung cancer. Vanadium toxicity is intensified by high dietary zinc, and is reduced by ascorbic acid, high protein, chromium, and manganese.

Niobium. Niobium is not toxic at physiologic levels. Cationic niobium is more toxic than are niobates following parenteral administration, because cationic niobium salts hydrolyze to form colloidal compounds. Hepatic and renal damage is the major toxic effect of niobium.

Tantalum. Tantalum salts are practically nontoxic when administered orally. Metallic tantalum is inert enough to be used as bone supports in surgery. Inhalation of Ta_2O_5 causes transient bronchitis and interstitial pneumonitis with hyperemia.

Group VI Metals

Among the metals of Group VI, toxicity is associated more with anions than with cations. The anions of the lighter metals, which are essential in the nutrition of mammals, are more toxic than the anions of the heavier members. If the toxicity of the cations alone were to be compared, the increase in the inherent toxicity could be related directly to increased electropositivity and atomic weight. The variable valence states exhibited by these lighter metals seem to increase their inherent toxicity. Among anions, those with lower valence states are more toxic than those with higher valence states (e.g., selenite is more toxic than selenate). Anionic hexavalent chromium salts, however, are more toxic than cationic trivalent chromium salts. The anionic salts of metals of subgroup A are more toxic than the corresponding salts of the metals of subgroup B. Anionic chromium salts are more toxic than either molybdenum or tungsten anionic salts.

The inherent toxicity of these metals decreases directly with atomic weight and electropositivity on the basis of LD_{50} values of both cationic and anionic salts; in the respective subgroups the salts of the lighter metals are the most toxic in their highest state of oxidation. The metals of subgroup VIB are less toxic than the metals of subgroup VIA, because at physiologic pH the salts of subgroup B metals are converted to insoluble polyacids through olation, while the anionic selenium and tellurium readily complex with biologically active thiol compounds. The toxicity of the individual metals decreases from Se to Cr^{3+}: $Se > Te > Cr^{6+} > Mo > W > Cr^{3+}$.

Subgroup VIA Metals

Selenium. Selenosis is caused by both organic and inorganic forms of selenium. Symptoms of acute selenium toxicity in man are nervousness, fever, vomiting, somnolence, lowering of blood pressure, tetanic and clonic spasms, and death due to respiratory failure. Chronic selenium toxicity symptoms are depression, garlic odor of breath, coated tongue, and marked pallor. Loss of nails and hair, hemolytic anemia, and damage to the kidneys, liver, and spleen are some of the other clinical symptoms. Selenium exhibits significant toxic effects in reproduction, calcification, and dental caries. Selenium is teratogenic and causes abnormal development of embryos in experimental and domestic animals. There are reports of teratogenic effects in humans, which include miscarriages and the birth of clubfooted infants to female laboratory technicians exposed to selenium. Selenium protects against cadmium- and arsenic-induced teratogenic malformations in animals, and human neonatal deaths declined with increased environmental selenium in epidemiologic studies. The carcinogenicity of selenium has yet to be established unequivocally in experimental animals, according to the rigid criteria associated with metallic carcinogens. Dietary selenate at 3 ppm levels in drinking water causes liver tumors in experimental rodents, but these liver tumors do not metastasize; slightly higher doses cause mortality with no tumor formation. Epidemiologic studies reveal an inverse relationship between human cancer mortality and environmental and blood selenium levels. Selenium toxicity is generally attributed to its interference with sulfur metabolism. Within physiologic limits, animals are able to maintain nontoxic levels of selenium by homeostatic excretory mechanisms. Lysine or fish meal partially suppresses the toxicity of dietary selenium in rats. Adequate dietary protein, linseed oil meal, organic sulfur, sulfate, or arsenic decrease the severity of selenium toxicity through excretion. Dietary cobalt decreases selenium retention. Selenium decreases the toxicity of metals such as mercury, thallium, cadmium, and tellurium; conversely, these metals increase selenium levels in tissues. Selenium is now considered to be essential. It is most toxic!

Tellurium. Metallic tellurium and TeO_2 are poorly absorbed from the mammalian digestive tract and from the sites of parenteral injections; tellurites and tellurates are readily absorbed. Retention of tellurium in soft tissues and in bone is high. Elemental tellurium and its salts can cross blood–brain and placental barriers. Excretion of tellurium is both fecal and urinary, and, to a small extent, in exhaled breath. The acute toxic effects of tellurium are suppression of sweat, nausea, somnolence, inflammation of gastric mucosa, intense hyperemia of internal organs, intestinal hemorrhage, and respiratory failure leading to death. Chronic tellurium intoxication symptoms are somnolence, garlic odor of breath, digestive disturbance, and growth depression. These symptoms are similar in both man and animals.

Excess tellurium and its compounds affect the functioning of renal, hepatic, brain, and nerve tissues. Rats with tellurium toxicity exhibit paralysis of the hind limbs due to segmental demyelination of the sciatic nerves and spinal roots. Excess tellurium fed to pregnant animals causes hydrocephalus in fetuses and neonates. The only form of tellurium detoxication known is the immobilization of elemental tellurium in an inert form as black pigment in soft tissues.

Subgroup VIB Metals

Chromium. Chromium (Cr^{3+}) is considered to be the least toxic among the essential trace metals on the basis of the essential-to-toxic ratio. The absorption of chromium salts from the digestive tract is low, since hexavalent anionic chromium is reduced to Cr^{3+} in the acid fluid of the stomach, and Cr^{3+} is poorly absorbed. Trivalent chromium and its salts do not cross the placental membrane, whereas the chromium-containing glucose tolerance factor is easily transferred.

Hexavalent anionic chromium is topically corrosive, and oral ingestion is toxic. In rabbits, large doses of chromate cause albuminuria with desquamated cells and renal hyperemia, fatty generation, and necrosis. Nephritis, anuria and extensive lesions in the kidney, and gastrointestinal ulceration are noticed in humans suffering from Cr^{6+} toxicity. Chromate contact dermatitis extends from a dry erythematous condition to eczema on the hands and other exposed parts. Inhalation of Cr^{6+} causes inflammation and ulceration of the nasal mucosa. Pulmonary carcinoma in humans 3–4 years after initial inhalation exposure is linked to Cr^{6+}; experimental evidence in animals is lacking. Intramuscular implantation of CrO_3 or $CaCrO_4$ in rats, however, causes a high percentage of malignant tumors.

Molybdenum. The severity of molybdenum toxicity depends on other nutrients and its chemical form and dose; increased protein intake, dietary copper, zinc, and organic sulfur (methionine) decrease the effects of molybdenum toxicity. Acute molybdenum toxicity causes severe gastrointestinal irritation with diarrhea, coma, and cardiac failure ending in death. Chronic molybdenum intoxication in animals causes anemia, anorexia, poor growth or loss in weight, male sterility due to testicular degeneration, poor conception and deficient lactation, teart syndrome, osteoporosis, and bone-joint abnormalities. Inhalation of soluble MoO_3 dusts causes bronchial and alveolar exudation. High levels of molybdate inhibit *in vivo* activities of ceruloplasmin and enzymes such as cytochrome oxidase, glutaminase, choline esterase, and sulfite oxidase. Excessive intake of dietary molybdenum interferes with calcium and phosphorus metabolism, and induces osteoporosis, lameness, bone-joint abnormalities, and connective tissue changes. Dietary supplements of Cu^{2+}, sulfur-containing amino acids, and sulfate inhibit molybdenum toxicity; hexavalent sulfate can displace molybdate by simple mass action and increase urinary molybdenum excretion.

Tungsten. Cationic and elemental tungsten are poorly absorbed, while anionic soluble tungstates are rapidly absorbed from the gastrointestinal tract. Intestinal and rumen microflora interfere with the absorption of tungsten. The excretion of tungstate is mainly urinary. Bone and soft tissues retain appreciable amounts of tungstate; however, laboratory rodents retain tungstate poorly in the liver and kidney. It is antagonistic to molybdenum.

Acute tungstate toxicity causes nervous prostration, diarrhea, respiratory paralysis, and death. Cationic tungsten is not as toxic as anionic tungstate. Chronic tungsten toxicity causes retarded growth. The production of molydenum-dependent enzymes is inhibited in neonates when pregnant animals are fed tungstate. In adult animals, dietary tungstate rapidly decreases the activities of xanthine oxidase and sulfite oxidase of liver, lung, and kidney.

Group VII Metals

Among the metals of subgroup VIIB, anions of the lighter member, manganese, are more toxic than are those of the other two metals, which have not been extensively investigated for their toxicity. Cationic rhenium is more toxic than is cationic manganese. Again, the general pattern of increased inherent toxicity with increased electropositivity and atomic weight is observable. At physiologic pH, however, the solubility of cationic rhenium salts is decreased, and the degree of hydrolysis to insoluble compounds is increased. Another interesting feature is the increased toxicity of the oxyacid anions; the higher the valency state, the greater is the toxicity of the oxyacids in Group VII metals.

Metals of subgroup VIIB appear to be the least toxic of the soluble heavy metal salts; the cationic salts are generally more toxic than are the anionic salts when administered parenterally. Within the subgroup, the toxicity associated with this group appears to increase directly with higher atomic weight and electropositivity. Although chemical toxicity data on technetium are not available, its toxicity could be predicted to be between manganese and rhenium.

Manganese. Chronic manganese toxicity in humans causes "manganism," affecting the central nervous system and involving psychic and neurologic disorders. Nephritis, cirrhosis of the liver, anorexia, muscular fatigue, and leukopenia are the symptoms of manganism. Massive feeding of Mn^{2+} to experimental animals retards growth, and causes calcium loss and poor absorption of iron, which leads to anemia, negative phosphorus balance, and rickets. Manganese toxicity is associated with renal degeneration due to increased DNAse and acid phosphatase and decreased ATPase activities. An efficient homeostatic mechanism prevents manganese accumulation in tissues, and toxic effects of manganese are not clearly manifested.

Technetium. Data on the chemical toxicity of technetium are not available, because stable isotopes of this artificial metal do not exist. Technetium accumulates in the lung, kidneys, liver, spleen, thyroid, and brain when administered parenterally in the form of organic chelate complexes. Its toxicity is probably between that of manganese and rhenium.

Rhenium. Rhenium exhibits low toxicity. Symptoms of acute rhenium toxicity in rats include sedation, abdominal irritation, severe ataxia, and tonic convulsions; cardiovascular collapse causes death. Cationic Re^{3+} is 10 times more toxic than the heptavalent ReO_{4-} anion.

Group VIII Metals

The metals of Group VIII differ from other groupings since they are subdivided, not into subgroups A and B, but into sets of horizontal triads. When these metals are divided in an unconventional manner into three vertical subgroups (iron, rubidium, osmium; cobalt, rhenium, iridium; nickel, paladium, platinum), the metals of the first and third subgroups are more toxic than those of the second subgroup. The lighter metals of Group VIII are essential; the soluble cationic salts and volatile compounds of these metals are more toxic than the heavier metals. However, the anionic complex soluble salts of the heavier group (e.g., chloroplatinates, osmic acids) are more toxic than the anionic complexes of the lighter metals. Thus, increased inherent toxicity concomitant with increased electropositivity and atomic weight can be speculated for the Group VIII metals. Cobalt, nickel, rhodium, and palladium are inherently carcinogenic.

Iron. Hemosiderosis may occur in man when acid food or drink is prepared in iron pots. Iron overload occurs in mammals due to a regulatory homeostasis failure that alters the absorptive capacity of the digestive tract. While the absorbed iron accumulates in the parenchymal cells, parenterally administered iron accumulates in reticuloendothelial cells. Hemosiderosis and hemochromatosis cause a series of heterogeneous disorders.

Symptoms of acute iron poisoning are increased respiration and pulse rates, with congestion of blood vessels leading to hypotension, pallor, and drowsiness in 6–8 hr. Prostration, coma, and finally death due to peripheral cardiac failure occur in 36 hr. Chronic iron poisoning results in hemorrhagic necrosis of the gastrointestinal tract, hepatotoxicosis, metabolic acidosis, greatly prolonged blood clotting time, and elevation of plasma levels of serotonin and histamine. Hepatic damage causes jaundice by raising the serum bilirubin level and inhibiting hepatic enzymes. Parenteral administration of an iron–dextran complex induces malignant tumors at the site of injection, although most other iron salts and dextran alone do not induce tumors. Iron oxides enhance the carcinogenic action of organic carcinogens such as benzpyrene, presumably becoming inert carriers of the carcinogens into healthy cells. Accumulation of particulate iron compounds in lysosomes appears to be the only specific detoxication mechanism for iron.

Cobalt. Intestinal absorption of cobalt increases in iron deficiency. Cobalt accumulates in the bodies of mammals, since cobalt absorption readily exceeds its excretion. The toxicity of cobalt compounds is greatly influenced by the mode of intake.

Symptoms of acute oral cobalt intoxication include diarrhea, paralysis of limbs, and lowering of blood pressure and body temperature prior to death. Congestion of all organs, focal hemorrhages in the serosa of the intestines, hemorrhages in the liver and adrenals, thickening of alveolar epithelium in lungs, and renal and pancreatic degeneration are reported in acute cobalt toxicity. Inhalation of metallic cobalt powder or salts, or intratracheal injection of cobalt salts, causes acute lung irritation, edema and hemorrhage, hyperplasia of bone marrow, massive pericardial effusion, and extensive loss of fluids from capillaries into the peritoneal cavity.

Symptoms of chronic cobalt poisoning include polycythemia, hyperplasia of bone marrow and thyroid gland, pericardial effusion, and damage to alpha cells of pancreas. Aerosols containing cobalt compounds cause pulmonary edema. Excess cobalt salts produce polycythemia in all species; the dosage depends on the species and the animal's condition. Polycythemia occurs in calves prior to rumen development, but not in mature animals. Hyperplasia in bone marrow produces release of immature erythrocytes and reticulocytes with impaired respiratory function. Heart tissue is susceptible to cobalt toxicity. This tissue depends on oxidative metabolism for its high energy requirements, and cobalt ions bind and inactivate sulfhydryl groups. Dietary cobalt compounds are not involved in carcinogenesis in mammals, but parenteral injection of cobalt metal powder, oxides, sulfides, and other compounds produces malignant tumors in experimental animals at the site of injection, in the thyroid gland, and in other organs. When injected, some insoluble cobalt compounds are phagocytized by macrophages and detoxified in an unknown manner.

Nickel. Among the three metals—iron, cobalt, and nickel—that share the same transport mechanism across the intestinal wall, nickel compounds are absorbed the slowest and in lower quantities than iron or cobalt compounds. Lungs retain inhaled nickel salts which are subsequently removed slowly. Removal of nickel salts from sites of parenteral injection is rapid. Nickel metal is less toxic than its salts.

Symptoms of acute nickel toxicity are severe gastroenteritis, tremor, choreal movements, and paralysis. Death occurs primarily from heart failure. Chronic toxicity causes degenerative changes in heart muscle and in brain, lung, liver, and kidney tissues. Constant contact with nickel alloys causes contact dermatitis. Nickel is carcinogenic. Inhalation of nickel salts and nickel carbonyl causes respiratory tract neoplasia and myocardial infarction. Chronic exposure to low levels of nickel carbonyl produces squamous metaplasia of the bronchial epithelium and carcinoma of the respiratory tract and lungs.

Growth retardation following nickel intoxication is due primarily to inhibition of enzymes, such as cytochrome oxidase, isocitric dehydrogenase, and malic dehydrogenase of the liver, heart, and kidneys. Reproductive capacity is impaired by reduced spermatogenesis and sperm motility. Nickel carbonyl is the most toxic of all nickel compounds, irrespective of mode of intake. Finely divided nickel or colloidal nickel causes sarcoma of bone, connective tissue, nerve tissue, and muscles.

Ruthenium. Ruthenium salts, which are somewhat rare, are considered to be slightly toxic. Ruthenium chelates are readily absorbed and rapidly excreted. A transient retention of ruthenium occurs in the kidney.

Rhodium. Rhodium salts are also considered to be slightly toxic; they are carcinogenic in experimental animals. Rhodium toxicity is attributed to its effects on the central nervous system.

Palladium. Palladium salts are poorly absorbed from the digestive tract, and cause black coloration at the sites of parenteral injection. High dilutions of colloidal palladium hydroxides cause hemolysis of erythrocytes; hence, instantaneous death follows intravenous injection of soluble palladium salts. Malignant tumors of the lung and decreased growth are other symptoms of chronic palladium intoxication.

Osmium. Severe irritation and black coloration are noted in the mucous membrane of the lungs and gastrointestinal tract when osmium tetroxide comes in contact with those tissues. Data on osmium toxicity through oral and parenteral routes are not readily available. Inhalation of osmium tetroxide vapor or its contact with eyes produces acute toxic symptoms. The cornea and sclera become opaque, ulcerated, and covered with a brown film. A purulent bronchopneumonia with discoloration of the epithelial lining of the bronchi occurs. Chronic inhalation of very low doses of osmic tetroxide vapor causes reduced erythrocyte and leukocyte counts, degeneration of renal epithelium, and acute conjunctivitis. The insolubility of osmium salts presages the low toxicity of this metal.

Iridium. There are no reports on iridium toxicity. Most iridium compounds are not water-soluble and, hence, are poorly absorbed. High iridium toxicity could be predicted for soluble iridium salts, however, since iridium is capable of coordinating with sulfur- and nitrogen-containing ligands.

Platinum. Platinum metal is biologically inert and is used as support for fractured bones. Acute toxic symptoms of parenterally administered platinum salts are severe epileptiform convulsions, coma, and death in experimental animals. Chronic toxicity causes respiratory and skin disorders. Complex platinum salts act on the nervous system; simple platinum salts affect the digestive tract. Platinum oxides and simple salts cause contact dermatitis. Platinosis is considered to be an allergic response to complex platinum salts. The acute toxicity of chloroplatinates via intravenous injection is caused by histamine release.

PERIODICITY AND TOXICITY

Chemical Toxicity on the Basis of Vertical Groups

Review of the inherent toxicity of metals, relevant to their position in the vertical groups of the periodic table and their physiologic behavior (the solubility of the metallic salts in tissues and fluids; their absorption, transport, and distribution in the tissues; and their excretion and the retention in mammals) reveals interesting relationships among inherent toxicity, electropositivity, and solubility. Inherent toxicity increases with increased electropositivity and solubility of the metallic cations in water and lipids. Greater toxicity is found when electropositivity is associated with the formation of covalent and coordinate covalent compounds than when associated with the formation of electrovalent compounds. The most highly electropositive of the metals are the alkali group metals. These metals form strong polar and electrovalent compounds; their salts ionize in aqueous solution. The alkali metals do not normally form stable coordinate or chelate compounds with biologic macromolecules. Metals of Groups II and III, which are less electropositive than subgroup IA metals, form stable polar and covalent bonds with biologic macromolecules; the stability depends on the electropositivity of the metals. The greater the stability, the greater the inherent toxicity; the higher the solubility of the heavy metal salts in biologic fluids, the greater their inherent toxicity. This generalization does not apply to the alkali metals. Lead, thallium, and mercury ions are highly toxic due to their solubility. This solubility allows transport to tissues, where chelation or complexation with macromolecules can cause damage. Salts of Group III metals possess high inherent toxicity; however, these salts are hydrolyzed to insoluble hydroxides or oxides at the tissue pH. The sparingly soluble nature of these hydroxides immobilizes the metals at the site of administration and renders them less toxic. If the cations of subgroup IIIA metals are kept hydrated and in solution by suitable buffers or chemical agents, these metals exhibit high inherent toxicity.

Within most vertical groups of the periodic chart, increased electropositivity and inherent toxicity are associated with increased atomic number or weight. The heavier metals in each group have the capacity to form irreversible and stable complexes with biologic macromolecules, which changes their conformation and biologic function; hence, these heavy metals are toxic. The lighter metals form reversible complexes with the macromolecules, which allows essential functions to proceed. Barium and magnesium of subgroup IIA or mercury and zinc of subgroup IIB are suitable illustrations. An exception to this generalization is beryllium, the lightest metal of subgroup IA. Essential metals are not toxic in physiologic doses, but are toxic at higher levels.

The inherent toxicity of heavy metals is enhanced if the solubility of salts of these metals is increased; however, the aqueous solubility of the simple salts of the metals in a group generally decreases progressively with increases in atomic weight. Hence, the heavier metals cannot exert their potential inherent toxicity due to poor solubility. Lead, thallium, and mercury salts are exceptions to these generalizations due to the greater solubility of their simple salts. The inherent toxicity of subgroup A metals is usually greater than that of subgroup B metals due to the difference in their electronic configuration, especially in the outermost orbitals; however, some metals of subgroup B are more toxic than metals of subgroup A.

Chemical Toxicity on the Basis of Horizontal Periods

A review of the inherent toxicity of metals on the basis of horizontal periods in the periodic table reveals a pattern of toxicity that increases with successive periods to the extent of solubility. Within a given period there is a general progressive decrease in the inherent toxicity of the metal cation from Groups I to IV, followed by a progressive increase in the toxicity of the metal anion (oxyacid exhibiting the maximum valency state) from Groups V to VII. Metals of Group VIII exhibit a similar inherent toxicity associated with both types of ions. The increased solubility of the alkali salts of the oxyacids of metals of Groups V–VII enhances this inherent toxicity. This generalization holds more for horizontal periods 3 and 4 than for periods 5 and 6 or the incomplete period 7. The electropositivity of metals increases progressively with successive horizontal periods.

The first horizontal period comprises hydrogen and helium; they are not toxic. Lithium and beryllium, the lightest members of subgroup IA and

IIA, respectively, are the two metals in the second period; these two metals possess greater inherent toxicity than metals that are essential nutrients for mammals.

Sodium, magnesium, and aluminum of the third period are comparatively nontoxic by oral administration, although cationic magnesium and aluminum salts are toxic when introduced parenterally into experimental animals. The metals potassium through selenium of the fourth horizontal period contain the first transitional series of elements. These metals generally form soluble salts, both anionic and cationic, and are nontoxic at physiologic levels. Some salts of these metals in lower valence states are practically nontoxic (less than 1 in the pT_{50} scale). Some metals, such as selenium, are more toxic in lower valence states than in higher valence states. Metals in Groups III–V in this horizontal series do not exert their inherent toxicity because their soluble simple cationic salts are hydrolyzed to insoluble metal hydroxides or oxides; hence, they are not easily absorbed from the digestive tract. These salts exhibit their inherent toxicity when introduced parenterally. Citrates of these metals can be toxic because they are stable enough for absorption. Subsequent metabolic breakdown of the citrates in the tissues releases the metal ions within the tissues.

Metals of the fifth, sixth, and seventh horizontal periods are more electropositive than are metals of the lower-numbered periods, and their salts are progressively less soluble than those of the lower periods. The greater electropositivity of the metals of the fifth horizontal period, extending from rubidium to tellurium, renders them more toxic than the metals of the fourth period. Within the fifth period, metals of subgroups IIIB, IVB, VB, and VIB, and Group VIII are of low inherent toxicity compared with the metals of subgroups IB, IIB, IIIA, IVA, VA, and VIA; this is due mainly to the insoluble nature of the simple salts and the hydrolysis of soluble salts to insoluble oxysalts of the second series. The sixth horizontal period comprises 30 metals, including the lanthanides, or rare earth metals. These metals are the most electropositive of the nonradioactive metals of all horizontal periods, and, hence, possess the greatest inherent toxicity. They are capable of forming cationic and anionic salts that form stable complexes with cellular macromolecules and active metabolites. Simple salts of cesium, barium, osmium, mercury, thallium, and lead are more soluble than the corresponding salts of lanthanum, hafnium, tantalum, tungsten, rhenium, iridium, platinum, and bismuth. Some of the anionic oxysalts of hafnium and tungsten, and salts such as aurothiomalates or chloroplatinates, are somewhat soluble in water and biologic fluids. The most soluble metal salts in this series are those of cesium and tellurium. The soluble salts of the metals in this period are the most toxic of all metal salts, with mercury and thallium salts at the top of the toxicity scale.

Analogous to mercury, thallium, and lead salts, the salts of other metals in this series should also be toxic; their insoluble nature and poor absorption from the digestive tract mask their potential inherent toxicity. Salts of these metals are highly toxic when administered parenterally, however, although few experimental data are available to substantiate this observation.

The last horizontal period of metals extends from francium to element number 104. This series includes most of the highly radioactive, unstable, and artifically made metals. The metals of this series should possess very high inherent toxicity, but it is impossible to assess their chemical toxicity due to their nonavailability in adequate quantities, their high radioactivity, and their low solubility. The retention of these metals in animal tissues, especially the actinides, renders them the most dangerous metals when considering both chemical and radiation toxicity.

An overall review of the toxicity of metals in horizontal periods reveals the following pattern of inherent toxicity. The toxicity associated with metal cations decreases progressively from subgroup IA to IIA and from subgroup IIIB to subgroup VIIB; it increases progressively from Group VIII to subgroups IB, IIB, IIIA, and IVA, and decreases from VA to VIA. The toxicity of anionic oxysalts of metals increases from subgroup IVB to VIIB, and then from subgroup VA to VIA. The metals of Groups I–III do not form anionic oxysalts. The foundations for these summary statements are reviewed in appropriate chapters of Volume 2, which consider each group of metals and their toxicity in mammals.

Integration

The biologic activity of metals follows those characteristics that determine their place in the periodic table (see p. 196). The lightest metals, lithium and beryllium, have such high charge to mass ratios that they are not incorporated into essential reactions of mammalian life. The essential functions of life definitely require light metals. Most of the essential metals occupy the third and fourth periods of the table. Sodium, potassium, magnesium, and calcium of subgroups IA and IIA are the essential macrometals; those belonging to the other groups are essential trace metals. Molybdenum is the lone trace metal in the fifth period. Physiologic amounts of the lighter metals react to maintain the dynamic equilibrium of life processes.

Stimulatory activity may be either beneficial or harmful. Cancer induction, a "harmful" stimulatory action, is a property associated with metals of the fourth period and the reactive metals of the fifth period, excepting metals of the vertical subgroups IA and IIA. Heavy metals that possess stimulatory capacity are found in both strong and weak electropositive groups of the periodic table. Strong electroactivity (or shifting of valence electrons) apparently overcomes a certain mass restriction, enabling heavy metals to retain chemical and biologic reactivity. Metals of the fifth and sixth periods that form soluble salts, such as cesium and thallium, are predicted to be stimulatory when tested in minute quantities. Most of the other metals in these two periods are too heavy and too unreactive to be either essential or stimulatory. Salts of most of these metals hydrolyze to insoluble oxyacids, oxides, or hydroxides under physiologic conditions. The natural and man-made metals heavier than bismuth are all radioactive; the chemical toxicity of these metals is difficult to study and largely unknown, but their bone-seeking capacity enhances their lethality.

The basis of metal toxicity in mammals is the reaction of the metal with an essential metabolite to prevent that metabolite from participation in the dynamic equilibrium of life. Excess metals of the fourth period compete to displace (other) essential metals in cellular reactions. Toxic heavy metals bind biologic macromolecules in irreversible conformations. Any physicochemical property of the atomic structure that solubilizes heavy metals provides a basis for increased toxicity. Such proclivity is noted in thallium and indium. These metals promise to become more notorious as their environmental impact is elucidated. These and the metals around them form a toxic cabal in the periodic table. Beryllium is uniquely toxic.

APPENDIX A

1969 International Atomic Weights, based on the assigned relative atomic mass of $^{12}C = 12$: The following values apply to elements as they exist in materials of terrestrial origin and to certain artificial elements.

Element	Symbol	Atomic number	Atomic weight	Element	Symbol	Atomic number	Atomic weight
actinium	Ac	89	(227)	fluorine	F	9	18.9984
aluminum	Al	13	26.9815	francium	Fr	87	(223)
americium	Am	95	(243)	gadolinium	Gd	64	157.2_5
antimony	Sb	51	121.7_5	gallium	Ga	31	69.72
argon	Ar	18	39.94_8	germanium	Ge	32	72.5_9
arsenic	As	33	74.9216	gold	Au	79	196.9665
astatine	At	85	(210)	hafnium	Hf	72	178.4_9
barium	Ba	56	137.33	hahnium	Ha	105	(260)
berkelium	Bk	97	(247)	helium	He	2	4.00260
beryllium	Be	4	9.01218	holmium	Ho	67	164.9303
bismuth	Bi	83	208.9806	hydrogen	H	1	1.008_0
boron	B	5	10.81	indium	In	49	114.82
bromine	Br	35	79.904	iodine	I	53	126.9045
cadmium	Cd	48	112.41	iridium	Ir	77	192.2_2
calcium	Ca	20	40.08	iron	Fe	26	55.84_7
californium	Cf	98	(249)	krypton	Kr	36	83.80
carbon	C	6	12.011	kurchatovium	Ku	104	(261)
cerium	Ce	58	140.12	lanthanum	La	57	138.905_5
cesium	Cs	55	132.9055	lawrencium	Lr	103	(257)
chlorine	Cl	17	35.453	lead	Pb	82	207.2
chromium	Cr	24	51.996	lithium	Li	3	6.94_1
cobalt	Co	27	58.9332	lutetium	Lu	71	174.97
copper	Cu	29	63.54_6	magnesium	Mg	12	24.305
curium	Cm	96	(245)	manganese	Mn	25	54.9380
dysprosium	Dy	66	162.5_0	mendelevium	Md	101	(256)
einsteinium	Es	99	(254)	mercury	Hg	80	200.5_9
erbium	Er	68	167.2_6	molybdenum	Mo	42	95.9_4
europium	Eu	63	151.96	neodymium	Nd	60	144.2_4
fermium	Fm	100	(255)	neon	Ne	10	20.17_9

(*Cont'd*)

1969 International Atomic Weights (Cont'd)

Element	Symbol	Atomic number	Atomic weight	Element	Symbol	Atomic number	Atomic weight
neptunium	Np	93	237.0482	selenium	Se	34	78.9_6
nickel	Ni	28	58.7_1	silicon	Si	14	28.0855
niobium	Nb	41	92.9064	silver	Ag	47	107.868
nitrogen	N	7	14.0067	sodium	Na	11	22.9898
nobelium	No	102	(254)	strontium	Sr	38	87.62
osmium	Os	76	190.2	sulfur	S	16	32.06
oxygen	O	8	15.999_4	tantalum	Ta	73	180.947_9
palladium	Pd	46	106.4	technetium	Tc	43	98.9062
phosphorus	P	15	30.9738	tellurium	Te	52	127.6_0
platinum	Pt	78	195.0_9	terbium	Tb	65	158.9254
plutonium	Pu	94	(244)	thallium	Tl	81	204.3_7
polonium	Po	84	(210)	thorium	Th	90	232.0381
potassium	K	19	39.0983	thulium	Tm	69	168.9342
praseodymium	Pr	59	140.9077	tin	Sn	50	118.6_9
promethium	Pm	61	(147)	titanium	Ti	22	47.9_0
protactinium	Pa	91	231.0359	tungsten	W	74	183.8_5
radium	Ra	88	226.0254	uranium	U	92	238.029
radon	Rn	86	(222)	vanadium	V	23	50.941_4
rhenium	Re	75	186.2	wolfram	W	74	183.8_5
rhodium	Rh	45	102.9055	xenon	Xe	54	131.30
rubidium	Rb	37	85.467_8	ytterbium	Yb	70	173.0_4
ruthenium	Ru	44	101.0_7	yttrium	Y	39	88.9059
rutherfordium	Rf	104	(261)	zinc	Zn	30	65.3_7
samarium	Sm	62	150.4	zirconium	Zr	40	91.22
scandium	Sc	21	44.9559				

APPENDIX A

PERIODIC TABLE

IA	IIA	IIIB	IVB	VB	VIB	VIIB	VIII			IB	IIB	IIIA	IVA	VA	VIA	VIIA	NOBLE GASES
1 H 1.0079†																1 H 1.0079†	2 He 4.00260
3 Li 6.941†	4 Be 9.01218											5 B 10.81	6 C 12.011	7 N 14.0067	8 O 15.9994†	9 F 18.99840	10 Ne 20.179†
11 Na 22.98977	12 Mg 24.305											13 Al 26.98154	14 Si 28.0855	15 P 30.97376	16 S 32.06	17 Cl 35.453	18 Ar 39.948†
19 K 39.0983	20 Ca 40.08	21 Sc 44.9559	22 Ti 47.90†	23 V 50.9414†	24 Cr 51.996	25 Mn 54.9380	26 Fe 55.847†	27 Co 58.9332	28 Ni 58.71†	29 Cu 63.546†	30 Zn 65.38	31 Ga 69.72	32 Ge 72.59†	33 As 74.9216	34 Se 78.96†	35 Br 79.904	36 Kr 83.80
37 Rb 85.4678†	38 Sr 87.62	39 Y 88.9059	40 Zr 91.22	41 Nb 92.9064	42 Mo 95.94†	43 Tc 98.9062	44 Ru 101.07†	45 Rh 102.9055	46 Pd 106.4	47 Ag 107.868	48 Cd 112.41	49 In 114.82	50 Sn 118.69†	51 Sb 121.75†	52 Te 127.60†	53 I 126.9045	54 Xe 131.30
55 Cs 132.9054	56 Ba 137.33	57 *La 138.9055†	72 Hf 178.49†	73 Ta 180.9479†	74 W 183.85†	75 Re 186.2	76 Os 190.2	77 Ir 192.22†	78 Pt 195.09†	79 Au 196.9665	80 Hg 200.59†	81 Tl 204.37†	82 Pb 207.2	83 Bi 208.9804	84 Po (210)	85 At (210)	86 Rn (222)
87 Fr (223)	88 Ra 226.0254	89 †Ac (227)	104 (260)	105 (260)													

*Lanthanoid Series

58 Ce 140.12	59 Pr 140.9077	60 Nd 144.24†	61 Pm (147)	62 Sm 150.4	63 Eu 151.96	64 Gd 157.25†	65 Tb 158.9254	66 Dy 162.50†	67 Ho 164.9304	68 Er 167.26†	69 Tm 168.9342	70 Yb 173.04†	71 Lu 174.97

†Actinoid Series

90 Th 232.0381	91 Pa 231.0359	92 U 238.029	93 Np 237.0482	94 Pu (244)	95 Am (243)	96 Cm (247)	97 Bk (247)	98 Cf (251)	99 Es (254)	100 Fm (257)	101 Md (258)	102 No (255)	103 Lr (256)

§The International Union for Pure and Applied Chemistry has not adopted official names or symbols for these elements.

†These weights are considered reliable to ± 3 in the last place. Other weights are reliable to ± 1 in the last place.

Atomic weights corrected to conform to the 1971 values of the Commission on Atomic Weights.

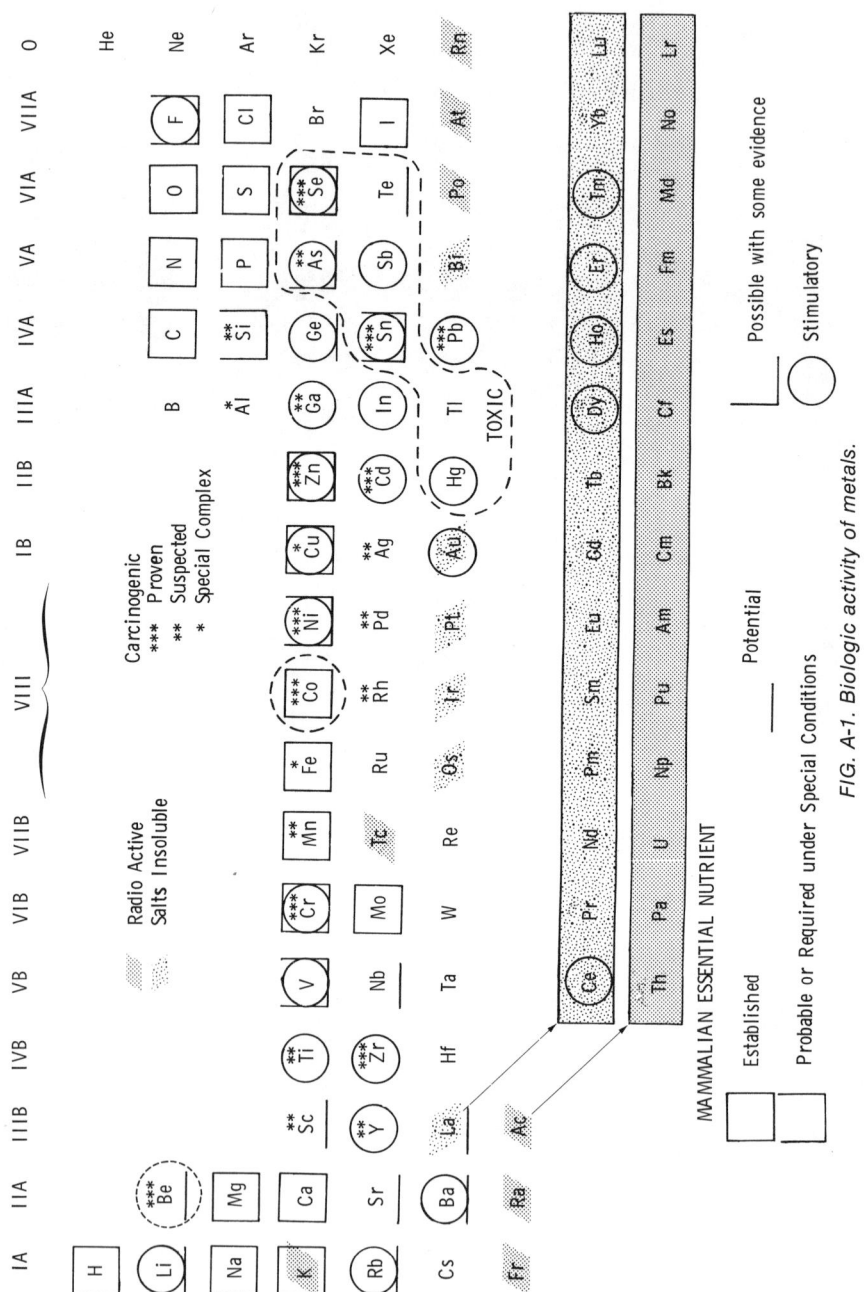

FIG. A-1. Biologic activity of metals.

APPENDIX B

Public concern for poisonous substances in our environment increases the need for better understanding, expression, and interpretation of the inherent toxicity of poisonous compounds. The usual standards, such as MLD, LD_{50}, and LD_{100}, express toxicity on a weight basis (mg material/kg body weight). This is of value under practical conditions. The molar concentration is meaningful for science but is confusing to those not technically oriented. Expression in terms of molarity (M/kg) is best for inherent toxicity and mechanisms of toxicity. The standards shown in Table 1-2 and 1-3 provide limited concepts of toxicity; the comparative toxicity is not fully expressed. The great differences in toxicity values between the extremely potent bacterial toxins and relatively nontoxic material, such as common inorganic salts, render these expressions still less useful.

A new expression for toxicity is introduced using the logarithm of the inverse function of the molar concentration. This provides a positive scale that is analogous to pH; the higher the number, the greater the toxicity. The inherent or potential toxicity of a compound is defined as the negative logarithm of the molar concentration for the stated conditions such as the species, mode of administration, time of day, LD value, etc.

$pT = -\log [T]$, where [T] is the number of moles of the toxic

compound per kg body weight, the molar concentration. Since the base of the logarithm is 10, each unit increase for pT indicates a ten-fold increase in toxicity. Toxicity data of some compounds in both conventional and pT expressions are given in Table 1-4. pT expressions are most useful in comparative toxicity and mechanism studies. Examples illustrate the effect of different expressions (see Table B-1). Comparison of the toxicity of two oxides reveal that BeO is ten times more toxic than the normal In_2O_3 on a compound weight basis and 2.5 times more toxic on a metal weight basis. However, BeO is only twice as toxic as In_2O_3 on a molar basis. This is true for either the compound or the metal; the pT expression indicates this.

TABLE B-1. Comparative Lethal Toxicity of Metal Salts in Rats
(Intraperitoneal Route)

Compound	Lethal dose	Toxicity expressed as			pT
		Compound mg/kg	Metal mg/kg	Metal mmole/kg	
BeO	MLD	54	19.4	2.15	2.67
In_2O_3	MLD	546	54.1	4.71	2.33
$BeCl_2$	LD_{50}	4.4	0.5	0.056	4.26
$ThCl_4$	LD_{50}	6.9	4.5	0.022	5.66
Be acetate	LD_{50}	317	22.4	2.49	2.6
Tl acetate	LD_{50}	29.6	23.0	0.113	3.94

Comparison of chlorides reveal that $BeCl_2$ shows 50% greater toxicity than $ThCl_4$ on a compound weight basis, and ten times greater toxicity on the metal weight basis. However, Th is clearly twice as toxic as Be on a molar basis, and this is indicated by the higher pT value. In the third example, Tl acetate is ten times more toxic than Be acetate on a compound weight basis, while the weight of the metals used indicates that they are equally toxic. However, comparison of the molar toxicity of the metals shows that Tl is clearly twenty times more toxic than Be, as indicated by the pT value.

Interconversions between [T] and pT are illustrated by examples. The acute ip LD_{50} toxicity of $BeCl_2$ is 4.4 mg/kg for rats. The molecular weight of $BeCl_2$ is 79.918. Convert mg to g and divide by the molecular weight in order to obtain the molar concentration:

4.4 mg = 0.0044 g
0.0044/79.918 = 0.00005506 = 5.5×10^{-5} mol
[T] = 5.5×10^{-5}

Calculate the pT value:

$$pT = -\log[T]$$

$$pT = \log \frac{1}{[T]} = \log \frac{1}{5.5 \times 10^{-5}}$$

From logarithm tables, the log of $5.5 \times 10^{-5} = 10^{0.7404} \times 10^{-5}$

$$pT = \log \frac{1}{10^{0.7404} \times 10^{-5}} = \log \frac{1}{10^{-4.2596}} \simeq 4.26$$

$$pT = 4.26$$

APPENDIX B

Determine the molar concentration from pT = 4.72

$$pT = 4.72 = \log 10^{4.72} = \log \frac{1}{10^{-4.72}}$$

$$= \log \frac{1}{10^{0.28} \times 10^{-5}}$$

Antilog 0.28 = 1.91; therefore

$$= \log \frac{1}{1.9 \times 10^{-5}} = -\log(1.9 \times 10^{-5})$$

$$[T] = 1.9 \times 10^{-5}$$

GLOSSARY

aboral: caudal or away from the mouth
accouterment: accessory trappings
actinide, actinon: a radioactive element resembling actinium, the lightest member of the series
active transport: movement of material by chemical energy (ATP) across a membrane against a concentration gradient
adenoma: a benign epithelial tumor of glandular origin
aerodynamic diameter: a dimension that is related to particulate performance in air flow and settling
aerosol: a colloid in which the continuous phase is a gas
agglutination: the clumping of cells or particles; the precipitation of a complex
aggression index: mortality data does not change with time following intoxication with toxic material
agribusiness: all business involved in the supply of natural food and textiles
alkali metal: any of the subgroup IA metals
alkaline earth metals: subgroup IIA metals
allergy: a hypersensitive reaction to environmental substances
allosteric effect: a conformational change brought about at a site other than the one directly affected
alluvial: material deposited or affected by running water
alopecia: excessive loss of hair
aluminum dextran: an aluminum derivative of a soluble starch
alveolus: a microscopic sac
amanitin(e): a poisonous alkaloid glycoside from certain mushrooms
amphibole: a metal silicate such as asbestos
amphoteric: capable of reacting in two ways—acidic or basic
anabolic: constructive metabolism
anaphylactic: a state of exaggerated allergic reaction
anaplastic: dedifferentiated

anion: an ion carrying a negative electric charge

anodic stripping voltametry: reverse polarography, giving a characteristic voltage for the metal that has been plated on the electrode; the total current at that characteristic voltage provides a quantitation of the metal

anorexia: loss of appetite for food

antimetabolite: an inhibitor that competes with a metabolite for an active site on a macromolecule or membrane

apoferritin: ferritin with little or no iron bound to it

arachnoid cavity: a cavity with a network of spaces

argyria: silver poisoning

argyrol: a registered name for a silver protein complex

aromatic polynuclear hydrocarbon: a steroid or other multiring aromatic carbon compound

asbestos: a fibrous magnesium or calcium silicate containing other metals

asbestosis: a form of lung disease caused by asbestos fiber inhalation

asthenia: weakness

astringent: an agent that causes contraction of blood vessels

atomic absorption: absorption of radiant energy by atoms with subsequent emission of radiation in spectral lines

axilla: the area and structures associated with the armpit

bactericidal: significantly destructive to bacteria

basement membrane: a delicate transparent membrane underlying the epithelium of mucous membranes and secreting glands

beryllosis: beryllium poisoning

bidentate: having two electron-donor atoms involved in a metal chelate

biliary obstruction: blockage of the bile ducts

bilirubin: a bile pigment derived from heme

biologic ligand: N-, O-, S-containing groups of biologic macromolecules that bind metal ions

biotransformation: a change in a molecule brought about through metabolism

bronchiole: one of the small, final branches of the respiratory tract

cancer: a cell growth or tumor that invades other tissues, usually to the detriment of the host

carbonyl: the CO radical or component of a compound; e.g., nickel carbonyl is $Ni(CO)_4$

carcinogen: a cancer-producing substance

carcinogenicity: the capacity to induce cancer

carcinoma: one of the two broad categories of cancer, a malignant growth of cells of epithelial origin

catabolic: processes of metabolism that degrade compounds

catechol: ortho-dihydroxybenzene

cathartic: an agent that causes rapid bowel elimination

GLOSSARY

cathepsin: intracellular protease
cation: an ion carrying a net positive electric charge
cell-mediated immunity: any of the cell–cell interactions that protect the body from foreign cells; any physical activity of cells in response to foreign material
ceruloplasmin: a glycoprotein that transports copper in the blood
charge redistribution: distribution of electrical charges in macromolecules by a shifting of electrons
chelate: a compound in which the metal can be incorporated through two or more donor atoms; metal chelates may have a ring structure
chelate, mixed: ring structure in which two or more dissimilar ligands are involved
choreal movements: ceaseless, jerky movements
chromosomal aberration: abnormality in the number or the constitution of chromosomes that may become expressed phenotypically
chronic toxicity: see *toxicity*
chronotoxicology: knowledge of the effect of diurnal or other time patterns on toxic effects
chrysolite: magnesium iron silicate, one form of asbestos
chyme: the semifluid digesta in the gastrointestinal tract
cilia: minute fingerlike projections of the cell, usually from a special area of the surface
cocarcinogen: an agent that enhances the carcinogenicity of a carcinogen
collagen: the supportive protein of connective tissue, skin, and skeletal tissues
colloid: finely dispersed matter in a dispersion medium
colloidoclastic: having the tendency to break a colloid into separate physical entities
coma: unconscious state
combining capacity: valency of metals, cations, and anions
conjuctivitis: inflammation of the conjunctiva, the delicate membrane that lines the eyelids
coordinate covalent bond: a chemical bond in which both shared electrons are provided from only one of the two atoms involved
coordinate bond: a chemical bond involving partial sharing of electrons
coordination number: the number of monovalent groups or radicals that could combine with a polyvalent ion through coordinate bonds
coprecipitation: the agglutination of an insoluble material by the precipitation of another complex
cornification: conversion of epithelium to heavy stratified squamous epithelium
corpus striatum: a component of the basal ganglia in the brain
covalent bond: a chemical bond between two atoms that share two electrons, one of which was supplied by each atom
croton oil: an oil used as a cathartic or vesicant
crypt: a compartment formed by a blind tube or pit leading to an open surface

cumulative toxicity: see *toxicity*
cycasin: a plant toxicant and oncogen
cytochrome: a special electron transfer hemoprotein, e.g., cytochrome C
decarboxylase: an enzyme that catalyzes the removal of the carboxyl group from a compound
decay curve: the time–radiation relationship of a radioactive substance; the decrease of any function with time
delayed toxicity: see *toxicity*
demulcent: a material that decreases irritation in a tissue
demyelinate: to destroy or remove the myelin sheath of nerves
denaturation: the destruction of the natural structure of a protein; active proteins, such as enzymes, are inactivated by denaturation
desferrioxamine: a potent iron-chelating agent
desmosome: an intercellular attachment site
desquamation: shedding of dead epidermis
detoxication: a reduction in toxicity, usually referring to a metabolic action
diameter, aerodynamic: size of particles as used in inhalation studies
di-associated: having two microbic species inoculated into a germfree animal to make a tribiotic system
digesta: partially digested food, e.g., chyme
digital tremors: involuntary trembling of the fingers
dinuclear complex: two nitrogen bases linked to a metal
dipole distance: the small distance that separates two electrically or magnetically charged particles of opposite sign
distal toxicity: see *toxicity*
dithiozon: diphenyl thiocarbozone, a chelating agent for heavy metals
diuretic: an agent that increases urine production
DMBA: dimethylbenzanthrocene, a carcinogen
DTPA: diethylenetriamine pentaacetic acid
dysarthria: imperfect speech, often due to damaged nerves
dysphagia: difficulty in swallowing
eccrine sweat glands: ordinary (exocrine) sweat glands
EDDHA: ethylenediamine-*bis*(1-hydroxy phenyl)acetic acid, a metal-chelating compound
edema: excess intercellular fluid
EDTA: a metal-chelating agent, ethylenediaminetetraacetic acid
effusion: a pouring forth
electrode potential: the potential difference between a metal and a solution of its water-soluble salt
electronegativity: the ability to gain electrons
electron microprobe: an electron beam source and detector for the detection of minute quantities of metals

GLOSSARY

electron mobility: movement or shifting of electrons within a molecule

electron polarizability: the absorption or removal of electrons from a molecule, causing polarization

electron orbitals: the electron energy levels assumed by electrons around the atomic nucleus.

electron transport system: the transference of electrons from one compound to another of increasing redox potential; in mitochondria of mammals, it is used to reduce oxygen in the formation of H_2O

electrophoretic mobility: the movement of an ion through a liquid in an electric field

electrophilic: having a strong attraction for electrons

electropositivity: the ability to lose electrons

electrovalent: descriptive of a chemical reaction between two atoms involving a transfer of electrons

element: an atom, one of the 110± distinct kinds of matter

embolism: the blockage of an artery or vein by an unusual body

emesis: vomiting

emetic: an agent that induces vomiting

emission spectroscopy: the determination of a specific atom by measurement of a characteristic light or ray emitted at a high temperature

emphysema: abnormal air accumulation

endocytosis: the engulfing of material with vacuolization of the cell membrane

endoplasmic reticulum: an ultramicroscopic system of membrane-bound cavities ramified throughout the cytoplasm of most cells

enteral: pertaining to the small intestine

enterohepatic system: the liver–small intestine interchange of material absorbed by the ileum, carried to the liver, and discharged into the intestine with the bile

enzyme: a protein catalyst

erythrocyte: red blood cell

essential nutrient: an element or compound that cannot be synthesized at a rate commensurate with the needs of the organism

exogenous: outside the cell or organism

facilitated diffusion: the "carrier"-enhanced flow of a given ion or compound across a membrane; the movement is always down a concentration gradient without expenditure of energy

fascia: fibrous tissue associated with internal organs

ferritin: a protein–iron complex that contains about 23% iron by weight

ferroxamine: a simple iron-chelating compound

fibroma: a tumor composed of connective tissue

fibrosarcoma: usually a malignant tumor, a sarcoma that contains, or is derived from, primitive connective tissue

fibrosis: formation (or degeneration) of fibrous tissue

filariasis: a disease state due to threadworm infection

flameless atomic absorption: a type of atomic absorption carried out with the material being heated in a tunnel-shaped furnace
fluorimetry: analysis that depends on the development or measure of fluorescence
follicle: a pouchlike cavity
foreign body carcinogen: a nonspecific carcinogen that depends on size, shape, and surface characteristics for its activity
gas–liquid chromatography (GLC): chromatography that incorporates gas flow over a liquid film for partition development
gavage: forced feeding into the stomach
genotype: the genetic composition of the individual
germfree: having no demonstrable viable microorganisms
gingivitis: inflammation of the mucous membranes of the gums
glomerular filtration: the filtration of liquid in the uriniferous tubules of the kidney
gluteal muscle: the major muscles of the buttocks
glycocalyx: the glycoprotein–polysaccharide extrusion of ciliated cells
glycoprotein: a protein that contains appreciable quantities of carbohydrate
goblet cell: an epithelial cell containing a globule of mucin
granulation: the formation of small masses of tissues
granuloma (pl granulomata): a tumor mass of granulation tissue
groups: vertical columns of the periodic table
hapten: a simple substance that can unite with a protein to form an antigen
hard acid: an acid fragment of a compound that holds its valence electrons firmly
hard base: a base fragment of a compound that holds its electrons firmly
HEDTA: hydroxyethylenediaminetetraacetic acid, a metal-chelating compound
hematinic: an agent that improves the quality of the blood
hemochromatosis: excess deposition of iron in tissues
hemosiderosis: abnormal storage of iron in tissues without tissue damage
histamine: a compound that causes dilation of blood vessels
histogenesis: the development of tissues from undifferentiated cells of the germ layer of the embryo
homeostasis: the tendency of the organism to maintain the normal internal environment
hormesis: the act of excitation
hormetic: stimulatory
hormetin: a stimulant
hormology: the knowledge of excitation; the study of stimulation
horny layer: the layer of cells in the epidermis that become hardened with fatty degeneration and dehydration
HSAB: hard and soft acids and bases; a theory of combining properties promulgated by Pearson (1963)
humoral immunity: that part of the defense that encompasses antibody production

hydrocephalus: a condition in which fluid accumulates in the head to cause enlargement, mental deterioration, and convulsions
hydroxamate: a compound derived from the substitution of a hydrogen from the amino group of hydroxylamine (NH_2OH)
hypercalcemia: an excess of calcium in the blood
hyperemia: excess blood
hypermagnesemia: an excess of magnesium in the blood
hyperplasia: an abnormal increase in number and/or size of cells in a tissue
hypophosphite: a salt of hypophosphorous acid, H_3PO_3
hypotension: abnormally low blood pressure
imino group: RC=NH
infarction: the formation of an infarct; the necrosis of tissue due to local deficiency of blood supply
inherent toxicity: toxicity due to the internal character of the material
interdigitation: an interlocking of fingerlike processes
interpolate: to extend information in accordance with known information
interstitial: between cells
interstitial fluid: the low protein plasmalike fluid that bathes cells
intoxication, metal: poisoning from a metal in any form
intradermal: within the dermis
intraperitoneal: within the peritoneum
intrathecal: within the sheath of the spinal cord
intrinsic toxicity: inherent toxicity
inunction: application (with rubbing) to the skin
ionic radius: effective radius of a charged ion
ionization potential: energy or work required (expressed in electron volts) to remove a given electron from its atomic orbit and place it at rest at an infinite distance
iron–dextran: a complex of iron with dextran, a soluble glucose polysaccharide
itai-itai: a neurologic disease associated with people exposed to mercury intoxication; first observed in Japan
jaundice: yellow color due to excess of bilirubin in tissues
kala-azar: a parasitic infection from *Leishmania donovani* that is often fatal
keratin: an insoluble protein that is the main component of hair, nails, and horns
lanthanide, lanthanoid, lanthanon: any of the series of elements called rare earths with characteristics similar to lanthanum, the lightest of the series
laser: a beam of monochromatic radiation (intense and nondivergent)
latent carcinogenesis: carcinogenesis that develops after an appreciable time has followed the administration of the agent
LD_1, LD_{50}, LD_{100}: the lethal dose for 1, 50 or 100% of the subjects to which the agent was administered

leishmaniasis: infection with *Leishmania*
leukemia: a persisting excess of white blood cells
leukocyte: any of the white blood cells or ameboid, poorly staining tissue cells
lieberkühn crypt: the glandular sac at the base of intestinal villi
ligand: an organic molecule that donates electrons to form coordinate covalent bonds with metal ions
liposarcoma: a malignant tumor derived from primitive lipoblastic cells
lymphosarcoma: a malignant growth of lymphoid tissue other than Hodgkin's disease
lysosome: a small cytoplasmic sac that contains hydrolytic enzymes
macromolecule: a polymeric compound with high molecular weight, usually protein, nucleic acid, carbohydrate, or a mixture of these with lipid
macrophage: an ameboid, phagocytic cell that is a normal constituent of most tissues
malignant tumor: an invasive, metastasizing, and progressive tumor that usually continues to grow to the detriment of the host
malignant neoplasm: a malignant tumor
manic psychosis: a disease characterized by emotional overreactions
megalopolophilic: attracted to urban environments
MDAF: toxicity value associated with a minimal dose, always fatal
MDNF: toxicity value associated with a maximal dose, not fatal
mesothelioma: a tumor formed from the epithelium that covers serous membranes
metabolism: the changes in molecules within living organisms, the sum of anabolism and catabolism
metal: any element that forms positive ions in solution
metallothionein: a protein isolated from the liver that is capable of binding toxic metals such as mercury and cadmium
metastasize: to spread from one tissue to another by cell transfer; the main feature of malignant tumors
microcalyx: the mucin extruded from the brush border
microcolloid: see *radiocolloid*
micron: 10^{-6} m; 1 micrometer; 1 μm
microvilli: the fingerlike protrusions from cells lining the villus, the brush border
milieu: the environment
millennia: thousands of years
minamata disease: mercury poisoning characterized by severe neurologic disorders
mineral: an inorganic homogeneous solid
MLD: minimum lethal dose
molal: moles per kilogram
molar: moles per liter
mole: the molecular weight in grams

monognotophoric: a host associated with only one microbic species
mucosa: a mucous membrane
multidentate: having more than two ligands involved in a metal chelate
myoblast: an embryonic cell capable of developing into a muscle cell
myocardia: heart muscle
myosarcoma: a malignant tumor derived from muscle tissue
nanogram: 10^{-9} g; 1 billionth of a gram; 1 ng
nature (of an organism): philosophically, the genetic makeup of the individual
necrosis: cell death, usually in a localized area
neoplasm: a new, abnormal growth
nephrotoxic: harmful to the kidney
neuraminic acid: sialic acid, a complex sugar
neuritis: painful nerve inflammation
neutron activation: the production of unstable isotopes by neutron bombardment of atoms
neutron flux: the density of neutrons in a neutron beam
nonspecific carcinogen: a surface carcinogen, one that depends for its activity more on physical size, shape, and polish than on chemical character
nature (of an organism): philosophically, the genetic makeup of the individual individual
occlusion: the hiding of a chemical by absorption or other means
olation: hydrolytic salt formation, e.g., BiOCl or Bi(OH)$_2$ Cl
oligodynamic: active in minute quantities; the olidogynamic action of heavy metals refers to the bactericidal activity of minute concentrations of heavy metal ions
oncogenic: carcinogenic
organelle: any discrete functioning unit of a cell
organogenesis: the development of organs
oxolation: addition of OH groups to products of the hydrolysis of metal salts
oxyacid: an acid such as perchloric or sulfuric, containing oxygen, hydrogen, and other electronegative atoms
oxysalt: a salt of an oxyacid
palliative: easing without curing
pallor: paleness
papillary: resembling or pertaining to a nipple
papilloma: an epithelial benign tumor
parameter: a variable that can be measured
parathion: an organic phosphorus insecticide
parenteral: outside the alimentary tract
passive diffusion: passage of material across a membrane with only the concentration gradient as the driving force
peptidase: an enzyme that catalyzes peptide hydrolysis
percutaneous: through the skin

period: a horizontal series in the periodic table
peritoneal cavity: the cavity formed within the peritoneum
peritoneal fluid: the fluid within the peritoneal cavity
peritoneum: the serous membrane that enfolds the viscera
peroxidase: an enzyme that catalyzes the oxidation of substrate in the presence of hydrogen peroxide
persorption: the intercellular entry of digesta into the interior of the villus
peyer's patches: lymphoid aggregates in the wall of the intestine
PH: $-\log [H^+]$, a measure of acidity
phagocytosis: the envelopment of solids by the cell membrane to form an intracellular vacuole
pharmacotoxic: pertaining to the relationship of toxicity to drug efficiency
pharyngeal: pertaining to the pharynx, the confluence of the nasal, oral, tracheal, and esophageal cavities
phosphodiester linkages: the linkages typified by cyclic AMP
phosphor: a phosphorescent compound or material
phosphorylase: an enzyme that catalyzes the hydrolysis of phosphate bonds
phylogenetic: pertaining to the development of the species or race
physicochemical properties: those properties that depend directly on atomic structure, as opposed to physical form
physiologic homeostasis: maintenance of the equilibrium of the organism by physiologic means
picogram: 10^{-12}g; 1 micromicrogram; pg
pinocytosis: taking fluid in through evagination of the cell membrane followed by vacuolization of the fluid
placental membrane: the semipermeable membrane that separates fetal from maternal blood
plaque: a flat, thin mass, e.g., a deposit on a tooth
pollutant: any noxious material that renders the environment unclean
polycythemia: an increase in red cells in the body
polydactyly: excess digits on hands or feet
polymeric: having repetitive units in the structure
polyneuritis: inflammation of many nerves
portal vein, portal blood: the vein or blood going from the intestine and spleen to the liver
ppb: parts per billion
ppm: parts per million
precancerous tissue: a conditioned tissue that will develop a tumor
primary carcinogen: a material that can induce cancer without undergoing change
primary site: the place of first contact
primary target tissue: the most sensitive tissue

prophylaxis: preventive treatment
prostigmin: trade name of neostigmine, a cholinergic drug
protease: a proteolytic enzyme
proximal toxicity: see *toxicity*
pT: −log [T], a measure of toxicity
pulmonary blood: blood within or from the lung
purgative: causing bowel evacuation
purulent: containing pus
quantum number: an index of the energy of a given electron level within an atom
quaternary ammonium hydroxide: ammonium hydroxide in which all four hydrogen atoms have been replaced by alkyl radicals
quaternary structure: the spatial relationships among separate subunits of a macromolecule
radioactive metal: a metal that gives off quantities of radiation that are significant from a health viewpoint
radiocolloid: a microcolloid; an especially small colloidal particle formed from a few molecules or ions that can be detected only through the use of radioactive tracers
radionuclide: a radioactive species of atom
recondite toxicity: see *toxicity*
refractory: unresponsive to treatment
replication: cell duplication or reproduction
RES: reticuloendothelial system
retention time: the period of time that the material is retained within the body or organ
reticuloendothelial system: the composite of all phagocytic cells in the body
retrobulbular: behind the eyeball
rhabdomyosarcoma: a malignant tumor of striated muscle
RNAase: an enzyme that catalyzes the hydrolysis of RNA
rotenone: an insecticide obtained from roots
salicylate: a salt of salicylic acid, *ortho*-hydroxy benzoic acid
sarcoma: a malignant growth of nonepithelial origin
saxitoxin: the poison from mussels and paralytic shellfish
scintillation counter: an instrument that converts a specific radiation into a numerical response
sebaceous gland: a gland that secretes fatty material
sebum: the secretion of sebaceous glands
secondary target tissue: tissues that are susceptible as secondary targets for toxic agents
sedimentation behavior: the characteristic centrifugal flow rate
sequester: to detach or hide from participation

serosal surface: that surface exposed to the blood supply
serotonin: a vasoconstrictor and an important neuroregulator in the brain
sigmoid: S-shaped
silica: silicon dioxide polymers
silicosis: a disease of the lungs caused by dust inhalation
soft acid: an acidic fragment of a compound incapable of holding its valence electrons firmly
soft base: a basic fragment of a compound not capable of holding its valence electrons firmly
spark-source mass spectrometry: a measurement by the mass spectograph, which uses an arc as the energy source to ionize the compound from the surface of the electrode
spindle cell fibrosarcoma: a sarcoma derived from fibroblasts in which the cells maintain their spindle shape
spirocheticidal: killing significant numbers of spirochetes
stoichiometric reaction: a reaction in which equivalent weights are involved
stomatitis: inflammation of the oral mucosa
stratum corneum: the outer stratum of dead cells of the epidermis
stratum germinativum: the innermost layer of reproducing cells of the epidermis
stratum granulosum: the granular layer of cells in the epidermis
stratum lucidum: the clear layers of cells beneath the stratum corneum
stratum mucosum: the thin layer of mucous cells
styptic: astringent or hemostatic
sublingual: beneath the tongue
sulfhydryl group: $-SH$, as in $R-SH$, thiol
suprapharyngeal: associated with the upper part of the pharnyx
swartzmann reaction: hemorrhagic necrosis at the site of injection of microbic filtrates
synergism: the working together to produce a greater action; antonym of antagonism
$[T]$: molar or molal concentration of a toxin or toxicant
teart: molybdenum toxicosis
teratogenesis: the production of physical fetal defects
teratogenicity: the activity that produces fetal anomalies
teratology: the knowledge of physical fetal anomalies
teratoma: a tumor containing fetal remains derived congenitally
terminal web: the clear area underlying the microvilli in a villus epithelial cell
tetrodotoxin: a toxin from the ovaries of the globe fish
thiol compounds: compounds containing $R-SH$; see *sulfhydryl group*
thiomalate: an ion or salt of mercaptosuccinic acid
thrombus: a blood clot or clump of cells that blocks a vessel
toxicant: a harmful agent other than the true protein toxins

toxicity: the degree to which a poison is harmful; different types are:

acute: toxicity in which critical symptoms develop quickly

chronic: toxicity resulting from exposure to small quantities for a long time

cumulative: toxicity in which effects are due to gradual accumulation of toxicant above a critical level

delayed: toxicity in which adaptation of the organism postpones the onset of symptoms

distal: toxicity in which the effects are observed long after administration of the toxicant

latent: equivalent to distal; a latent carcinogen may cause cancer 6–12 months later in mice or 10–20 years later in humans

proximal: toxicity in which the effects are shown immediately on administration of the toxicant

recondite: toxicity in which subtle effects follow prolonged exposure to threshold amounts of relatively harmless compounds

toxicosis: the morbid condition induced by a poison

toxin: a protein poison

transcription: the copying of a message, as the use of DNA to form RNA

transferrin: a serum β-globulin that binds and transports iron

transfollicular: pertaining to movement across follicles

transition metal: a metal with variable charges resulting from electrons occurring in varying orbits, e.g., iron

transit time: the time required for material to travel from one destination to another

translation: changing the information base, as the use of messenger RNA to dictate the sequence of amino acids in peptides

transuranium: any of those elements with atomic weights greater than uranium

trypanosome: a parasitic protozoan of the genus *trypanosoma*

tubular reabsorption: the recovery of filtrate during urine formation

tumor: an unusual growth of cells

unidentate ligand: one of the single ligands from each organic molecule involved in formation of a metal chelate

urethritis: inflammation of the urethra

X-ray fluoresence: fluorescence produced by exposing the material to radiation

ZEP: zero equivalence point

zero tolerance: detection of none of the material of interest in the sample

zymogen activation: conversion of a proenzyme into an active form

REFERENCES

Adam, M., and Kuhn, K. 1968. Investigation on the reaction of metals with collagen *in vivo*, *Europ. J. Biochem.* 3:407–410.
Agarwal, R. P., and Feldman, I. 1968. Chelation of uranyl ions by adenine nucleotides. II. Magnetic resonance investigation of the uranyl nitrate—adenosine 5 phosphate in D_2O at alkaline pH, *J. Am. Chem. Soc.* 90:6635–6639.
Agnese, T., Veris, B., and Stantolini, B. 1959. Incidence of lung cancer in relation to occupation in Genoa, *Igiene Mod.* 52:149–160.
Ahearn, A. J., 1972. *Trace Analysis by Mass Spectrometry*, Academic Press, New York.
Aikawa, J. A. 1971. *The Relationship of Magnesium to Disease in Domestic Animals and in Humans*, C. C. Thomas, Springfield, Illinois.
Albert, R. E., Lippmann, M., Spiegelman, J., Liuzz, A., and Nelson, N. 1967. The deposition and clearance of radioactive particles in the human lung, *Arch. Environ. Health.* 14:10–15.
Alexander, P., and Horning, H. S. 1958. *Ciba Foundation on Carcinogenesis: Mechanism of Action*. Little, Brown and Co., Boston.
Altman, P. L., and Dittmer, D. S. 1974. *Biology Data Book*, Vol. III. Federation of American Societies for Experimental Biology, Bethesda, Maryland.
Anderson, C. A. 1967. An introduction to the electronprobe microanalyzer and its applications to biochemistry, in: *Methods of Biochemical Analysis,* Vol. XV, Glick, D. (ed.), Interscience, New York.
Anke, M., Hennig, A., Schneider, H. J., Ludke, H., Von Cagern, W., and Schlegal, H. 1970. The interrelationships between cadmium, zinc, copper, iron in metabolisms of hens, ruminants and man, in: *Trace Elements Metabolism in Animals,* Mills, C. F. (ed.), Livingstone, London.
Anon. 1974. EPA broadens its definition of carcinogen. *C and E News 1974*:13.
Anon. 1974. The determination of all detectable elements in the aquatic plants of Linsley Pond and Cedar Lake (North Branford, Connecticut) by x-ray emission and optical emission spectroscopy, *Appl. Spectrosc.* 27:5.
Ansari, M. S., Miller, W. J., Lassiter, J. W., Neathery, M. W., and Gentry, R. P. 1975. Effects of high dietary zinc on zinc metabolism and adaptation in rats, *Proc. Soc. Exp. Biol. Med.* 150:534–536.
Aoki, F. Y., and Ruedy, J. 1971. Severe lithium intoxications: Management without dialysis and report of a possible teratogenic effect of lithium, *Can. Md. Ass. J.* 105:847.
Arena, J. M. 1974. *Poisoning: Toxicology, Symptoms, Treatments,* 3rd ed., American Lecture Series, No. 903, *American Lectures in Living Chemistry,* C. C. Thomas, Springfield, Illinois.

Arnold, M., and Sasse, D. 1961. Quantitative and histochemical analysis of Cu, Zn, and Fe in spontaneous and induced primary tumors of rats, *Cancer Res. 21*:761–766.
Arvela, P., and Karki, N. T. 1970. The effect of cerium on drug metabolizing activity in rat liver, *Acta Pharmacol. Toxicol. Suppl. (KBH)28*:36.
Aue, W. A., and Hill, H. H., Jr. 1972. A hydrogen-rich flame ionization detector sensitive to metals, *J. Chromatogr. 74*:319–324.
Autian, J. 1975. Toxicological problems and untoward effects from plastic devices used in medical applications, in: Hayes, W. J., Jr. (ed.), *Essays in Toxicology*, Vol. VI, Academic Press, New York.
Baetjer, A. M. 1956. Relation of chromium to health, in: *Chromium*, Vol. I: *Chemistry of Chromium and its Compounds*, Udy, M. J. (ed.), American Chemical Society Monograph, No. 132, pp. 76–104.
Baetjer, A. M., Lowney, J. F., Steffee, H., and Budacz, V. 1959. The effect of chromium on the incidence of lung tumors in mice and rats, *Arch. Ind. Health 20*:124–135.
Bair, W. J., and Thompson, R. C. 1974. Plutonium–biochemical research, *Science 183*:715–722.
Ball, R. A., Van Gelder, G., and Green, J. W. 1970. Neoplastic sequelae following subcutaneous implantation of mice with rare earth metals, *Proc. Soc. Exp. Biol. Med. 135*:462.
Bamann, E., Trapmann, H., and Fischler, F. 1954. Verhalten und Spezifität von Cer und Lanthan also Phosphatase—Modelle gegenüber Nucleinsäuren und Mononucleotiden, *Biochem. Z . 326*:89–96.
Barber, R. S., Braude, R., Mitchell, K. G., and Rook, J. A. F. 1956. Further studies on antibiotic and copper supplements for fattening pigs, *Proc. Nutr. Soc. 15*:IX, X.
Barnett, R. J. 1959. The demonstration with the electron microscope of the end products of histochemical reactions in relation to the fine structure of cells, *Exp. Cell Res.* (suppl. 7):65–89.
Bates, D. V. 1972. Air pollutants and the human lung, *Am. Rev. Respir. Dis. 105*:1–13.
Beckley, J. H., 1965. Comparative eye testing: Man vs. animals, *Toxic Appl. Pharmacol. 7*:93–101.
Beer, M., Stein, S., Carmalt, D., and Mohlhenrich, K. H. 1966. Determination of base sequences in nucleic acid with the electron microscope. V. The thymine specific reactions of OsO_4 with DNA and its components, *Biochemistry 5*:2283–2288.
Behrens, B., and Kärber, G. 1935. Wie sind Reihenversuche für biologische Auswertungen am zweckmässigsten anzuordnen?, *Arch. Exp. Path. Pharmakol, 177*:379–388.
Bessman, S. P., Rubin, M., and Leikin, S. 1954. The treatment of lead encephalopathy—a method for the removal of lead during the acute stage, *Pediatrics 14*:201–208.
Bettley, F. R. 1965. The influence of detergents and surfactants on epidermal permeability, *Brit.J. Derm. 77*:98–100.
Bienvenu, P., Noire., and Cier, A. 1963. The comparative general toxicity of metal ions: Relation with the periodic classification, *C.R. Acad. Sci. 21*:1043–1044.
Bingham, R. A., and Elliott, R. M. 1971. Accuracy of analysis by electrical detection in spark source mass spectrometry, *Anal. Chem. 43*:43–45.
Bischoff, F., and Bryson, G. 1964. Carcinogenesis through solid-state surfaces, in: Homburger, F. (ed.), *Progress in Experimental Tumor Research*, Vol. V, Karger, Basel and Hafner Publishing Co., New York, pp. 85–133.
Blamberg, D. L., Blackwood, U. B., Supplee, W. C., and Combs, C. F. 1960. Effect of zinc deficiency in hens on hatchability and embryonic development, *Proc. Soc. Exp. Biol. Med. 104*:217–220.
Blejer, H., and Wagner, W. 1976. Case study for: inorganic arsenic–ambient level approach to the control of occupational cancerigenic exposures, *Ann. N.Y. Acad. Sci. 271*:179–186.

REFERENCES

Bois, P. 1964. Tumor of the thymus in magnesium-deficient rats, *Nature 204*:1316.
Bonnell, J. A. 1955. Emphysema and proteinuria in men casting copper–cadmium alloys, *Br. J. Ind. Med. 12*:181–197.
Boulos, B. M., Carnow, B., Naik, N., Bederka, J. P., Kauffman, R. F., and Azarnoff, D. L. 1973. Placental transfer of lithium and environmental toxicants and their effects on the newborn, *Fed. Proc. 32(3)*:745.
Boutwell, R. K. 1963. A carcinogenicity evaluation of potassium arsenite and arsanilic acid, *J. Agric. Food Chem. 11*:381–385.
Boyd, E. M. 1972a. *Predictive Toxicology,* John Wright and sons, Bristol.
Boyd, E. M. 1972b. *Respiratory Tract Fluid,* C. C. Thomas, Springfield, Illinois.
Boyland, E. 1968. The correlation of experimental carcinogenesis and cancer in man, in: *Progress in Experimental Tumor Research,* Vol. II., Homburger, F. (ed.), Karger, Basel and Hafner Publishing Co., New York, pp. 222–234.
Boyland, E., Dukes, C. E., Grover, P. L., and Mitchley, B. C. V. 1962. The induction of renal tumors by feeding lead acetate to rats, *Br. J. Cancer 16*:283–288.
Brand, K. G., Buoen, L. C., and Brand, I. 1967. Malignant transformation and maturation in non-dividing cells during polymer tumorigenesis, *Proc. Soc. Exp. Biol. Med. N.Y. 124*:675–678.
Braude, R. 1965. Copper as a growth stimulant in pigs (cuprum pro pecunia), *Symposium on Copper's Role in Plant and Animal Life,* Vienna, Austria, pp. 55–66.
Braun, P., Guillerm, J., Pierson, B., and Sadoul, P. 1960. Bronchial cancer in workers in iron mines, *Rev. Med. Nancy 85*:702–708.
Braun, A., and Kern, V. 1968. Attempt to induce tumors by subcutaneous and intraperitoneal administration of ferridextran (spofa), *Neoplasma 15*:21–27.
Brookes, E. J., Tousimis, A. J., and Birkis, L. S. 1962. The distribution of calcium in the epiphyseal cartilage of the rat tibia measured with the electron probe X-ray microanalyzer, *J. Ultrastruct. Res. 7*:56–58.
Brown, R., Powers, P., and Wolstenholme, W. A. 1971. Computerized recording and interpretation of spark-source mass spectra, *Anal. Chem. 43*:1079–1083.
Bryan, S. E., and Frieden, E. 1967. Interaction of copper(II) with deoxyribonucleic acid below 30 degrees, *Biochemistry 6*:2728–2734.
Bryson, G., and Bischoff, F. 1969. The limitations of safety testing, *Prog. Exp. Tumor Res. 11*:100–133.
Butt, E. M. 1960. Trace metals in health and disease, Air Pollution Medical Conference, San Francisco, California.
Cairns, J. 1975. The cancer problem, *Sci.Am. 233*:64–78.
Calley, J. R. T. 1974. Smoking harm to your children, *Sci. News 106*:376.
Canada Food and Drug Protectorate. 1965. Guide for completing preclinical submission on investigational drugs, Department of National Health and Welfare, Ottawa.
Cannon, W. B. 1929. Organization for physiological homeostasis, *Phsyiol. Rev. 9*:399–431.
Carter, R. L., Roe, F. J. C., and Peto, R. 1971. Tumor induction by plastic films: Attempts to correlate carcinogenic activity with certain physicochemical properties of the implant, *J. Nat. Cancer Inst. 46*:1277–1283.
Casarett, L. J. 1972. The vital sacs: Alveolar clearance mechanisms in inhalation toxicology, in: Hayes, W. J., Jr. (ed.), *Essays in Toxicology,* Vol. III, Academic Press, New York.
Casarett, L. J., and Doull, J., (eds.). 1975. *Toxicology, The Basic Science of Poisons,* MacMillan Publishing Co., New York.
Castaing, R. 1951. Applications des sondes electroniques à une method d'analyse ponctuelle chimique et cristallographique, Ph.D. Thesis, University of Paris, ONERA, pub. 55.
C. D. C. 1975. Acute copper poisoning, *Morb. Mort. Week. Rep. Mar. 15*:99.

Cember, H., Hatch, T. F., Watson, J. A., Grucci, T., and Bell, P. 1956. The elimination of radioactive barium sulfate particles from the lung, *A.M.A. Arch. Ind. Health 13*:170–176.

Chahovitch, X. 1955. The action of Zn on the growth of experimental tumors incited by carcinogens, *Glas. Srp. Akad. Nauka, Umet. Od. Med. Nauka 215*:143–146.

Chandler, J. A. 1971. Ion beam analysis, *Am. Lab. 71*:50.

Christensen, H. E. and Luginbyhl, T. T. 1974. *Toxic Substances List,* U.S. Government Printing Office, Washington.

Clayson, D. B. 1962. *Chemical Carcinogenesis,* Little, Brown and Co., Boston.

Clemmens, E. T., Stevens, C. E., and Southworth, M. 1975. Sites of organic acid production and pattern of digesta movements in the gastrointestinal tract of the swine, *J. Nutr. 105*:759–778.

Cogburn, L. A., Parkhurst, C. R., and Thaxton, P. 1973. Relationship of mercury to reproductive efficiency and mating behavior in Japanese quail, Ann Meeting, Ass. of Southern Agric. Workers, Atlanta.

Coldwell, B. B., Solomanraj, G., Boyd, E. M., Jantz, J., and Morrison, A. B. 1969. The effect of dosage form and route of administration on the absorption and excretion of acetylsalicylic acid in man, *Clin. Toxicol. 2*:111–127.

Coogan, P. S. 1973. Lead induced renal carcinoma, *Proc. Inst. Med. Chic. 29*:309.

Cowgill, U. M. 1972. Selective detection of organometallics in gas chromatigraphic effluents by flame photometry, *J. Chromatogr. 74*:311.

Crane, R. K. 1975. 15 years struggle with the brush border, in: Csaky, T. Z. (ed.), *Intestinal Absorption and Malabsorption,* Raven Press, New York, pp. 127–142.

Creamer, B. 1974. *The Small Intestine,* Wm. Heinemann Medical Books, London.

Csaky, T. Z., and Autenrieth, B. 1975. Transcellular and intercellular intestinal transport, in: Csaky, T. Z. (ed.), *Intestinal Absorption and Malabsorption,* Raven Press, New York, pp. 177–186.

Currie, A. M. 1947. The role of arsenic in carcinogenesis, *Br. Med. Bull 4*:402.

Daniel, M. R. 1966. Strain differences in the response of rats to injection of nickel sulfide, *Br. J. Cancer 20*:866–895.

Danieli, J. F., and Davis, J. T. 1951. *Advances in Enzymology,* Vol. 11, Interscience Publishers, New York.

Dautreband, L. 1962. *Microaerosols,* Academic Press, New York.

Davenport, H. W. 1961. *Physiology of the Digestive Tract,* Yearbook Medical Publishers, Chicago.

Davies, C. N. 1961. *Inhaled Particles and Vapors,* Pergamon Press, New York.

Davson, H., and Danielli, J. 1952. *The Permeability of Natural Membranes,* 2nd ed., Cambridge University Press.

Dixon, J. R., Lowe, D. B., Richards, D. E., and Stockinger, H. E., 1969. The role of trace metals in chemical carcinogenesis: Asbestos cancers, in: Hemphill, D. D. (ed.), *Trace Substances in Environmental Health,* Vol. II, University of Missouri Press, Columbia, pp. 141–159.

Doll, R. 1958. Cancer of the lung and nose in nickel workers, *Br. J. Ind. Health 15*:217–223.

Domonkos, A. N. 1971. *Andrew's Diseases of the Skin,* W. B. Saunders Co., Philadelphia.

Dorland, W. A. N. 1974. *Dorland's Illustrated Medical Dictionary,* 25th ed., W. B. Saunders Co., Philadelphia.

Dove, W. F., and Davidson, N. 1962. Cationic effects on the denaturation of deoxyribonucleic acid, *J. Mol. Biol. 5*:467–478.

Dreyfus, J. R., 1936. Lungencarcinom bei Geschwistern nach Inhalation von eisenoxydhaltigem Staub in der Jugend, *Z. klin. Med. 130*:256–260.

Druckery, H., Hamperl, H., and Schmahl, D. 1957. Cancerogene Wirkung von metallischem Quecksilber nach intraperitonealer Gabe bei Ratten, *Z. Krebsforsch. 61*:511.

REFERENCES

Duckett, S. 1972. Teratogenesis caused by tellurium, *Ann. N.Y. Acad. Sci. 192*:220–226.
Dygert, H. P., LaBelle, C. W., Laskin, S., Pozzani, U. C., Roberts, G. F., Jr., and Stokinger, H. E. 1949. Toxicity following inhalation, in: Voegtlin, C. F., and Hodge, H. C. (eds.), *Pharmacology and Toxicology of Uranium Compounds*, McGraw-Hill Book Co., New York.
Eger, E. I. 1974. Uptake, distribution and elimination of inhaled anaesthetics, in: Scurr, C., and Feldman, S. (eds.), *Scientific Foundations of Anaesthesia*, Wm. Heinemann Medical Books, London.
Eichorn, G. L. 1973. Complexes of polynucleotides and nucleic acids, in: Eichorn, G. L. (ed.), *Inorganic Biochemistry*, pp. 1210–1243.
Eichorn, G. L., and Butzow, J. J., 1965. Interactions of metal ions with polynucleotides and related compounds. III. Degradation of ribonucleotides by lanthanum ions, *Biopolymers 3*:79–94.
Eick, J. D., Caul, H. J., Smith, D. L., and Rasberry, S. D. 1967. Analysis of gold and platinum group alloys by X-ray emission with corrections for interelement effects, *Appl. Spectrosc. 21*:324–328.
Eylar, E. H., Madoff, M. A., and Brody, O. V. 1962. The contribution of sialic acid to the surface charge of the erythrocyte, *J. Biol. Chem. 237*:1992–1998.
Fairchild, E. J., and Stockinger, H. E. 1958. Toxicologic studies on organic sulfur compounds. 1. Acute toxicology of some aliphatic and aromatic thiols (mercaptans), *J. Ind. Hyg. 19*:171–188.
Fare, G. 1965. Copper acetate as an accelerator in mouse skin carcinogenesis by 9,10-dimethyl-1,2-benzanthracene, *Experientia 21*:415–416.
Feldman, I., Jones, J., and Cross, R. 1967. Chelation of uranyl ions by adenine nucleotides, *J. Am. Chem. Soc. 89*:49–53.
Ferm, V. H., and Carpenter, S. J. 1967. Teratogenic effect of cadmium and its inhibition by zinc, *Nature (London) 216*:1123.
Ferm, V. H., and Carpenter, S. J. 1968. The relationship of cadmium and zinc in experimental mammalian teratogenesis, *Lab. Invest. 18*:429–434.
Ferm, V. H., and Carpenter, S. J. 1970. Teratogenic and embryopathic effects of indium, gallium and germanium, *Toxicol. Appl. Pharmacol. 16*:166–170.
Fielding, J. 1962. Sarcoma induction by iron carbohydrate complexes, *Br. Med. J. 1962*:1800–1804.
Fishbein, L. 1975. Identification of carcinogenic, mutagenic and teratogenic substances in the environment, *Environ. Qual. Saf. 4*:200–225.
Fisher, V. 1931. Intestinal absorption of viable yeast, *Proc. Soc. Exp. Biol. Med. 28*:948–951.
Flinn, R. H., Neal, P. A., Reinhart, W. H., and Dallavilk, J. M. 1940. Chronic manganese poisoning in an ore-crushing mill, *U.S. Publ. Bull. No. 247*, U.S. Govt. Printing Office, Washington, D.C.
Forth, W., and Rummel, W. 1973. Iron absorption, *Physiol. Rev. 53*:724–792.
Frankel, R. S., and Aitken, D. W. 1970. Energy dispersive X-ray emission spectroscopy, *Appl. Spectrosc. 24*:557–566.
Freeman, J. A., Geer, J. C., and Lillie, R. D. 1971. Fine structure of intestinal epithelial cell in iron absorption, in: Skoryna, S. C., and Waldron-Edwards, D. (eds.). *Intestinal Absorption of Metal Ions, Trace Elements and Radionuclides*, Pergamon Press, Oxford and New York.
Frieden, C. 1958. The dissociation of glutamic dehydrogenase by reduced diphosphopyridine nucleotide, *Biochem. Biophys. Acta 27*:431–432.
Frost, D. V. 1967. Arsenicals in biology—retrospect and prospect, *Fed. Proc. 26*:194–208.
Frost, D. V. 1970. Tolerances for arsenic and selenium, a psychodynamic problem. *World Rev. Pest Control 9*:6.

Frost, D. V. 1972. The two faces of selenium—Can selenophobia be cured?, *CRC Critical Review in Toxicology 1*(4):467–514.
Furst, A., and Haro, R. T. 1969a. A survey of metal carcinogenesis, *Prog. Exp. Tumor Res. 12*:102–133.
Furst, A., and Haro, R. T. 1969b. Possible mechanisms of metal ion carcinogenesis, in: Bergman, E. D., and Pullman, B. (eds.), *Physico-chemical Mechanisms of Carcinogenesis: Proceedings of an International Symposium Held in Jerusalem, 21 - 25 October, 1968*, Israel Academy of Sciences and Humanities, Jerusalem, p. 310.
Fuwa, K., Wacker, W. E. C., Druyon, R., Bartholomay, A. F., and Vallee, B. L. 1960. Nucleic acid and metals. II. Transition metals as determinants of the conformation of ribonucleic acids, *Proc. Nat. Acad. Sci. 46*:1298–1307.
Gamsu, G., Weintraub, R. M., and Nadel, J. A. 1973. Clearance of tantalum from airways of different caliber in man evaluated by a roentgenographic method, *Am. Rev. Respir. Dis. 107*:214–224.
Gavrilova, L. P., Ivanov, D. A., and Spirin, A. S. 1966. Studies on the structure of ribosomes. III. Stepwise unfolding of the 50 s particles without loss of ribosomal protein, *J. Mol. Biol. 16*:473–489.
Gibney, L. 1975. EPA broadens approach to pesticide decisions, *Chemical and Engineering News 53*:15–16.
Gilman, J. P. W. 1966. Muscle tumorigenesis, in: Begg, R. W., Leblond, C. P., Nobel, R. L., Rossiter, R. J., Taylor, R. M., and Wallace, S. E. (eds.), *Canadian Cancer Conference VI*, Pergamon Press, New York, p. 209.
Gilman, J. P. W., and Ruckenbauer, G. M. 1962. Metal carcinogenesis: Observations on the carcinogenicity of a refinery dust cobalt oxide and colloidal thorium oxide, *Cancer Res. 22*:152–157.
Gilman, P. F. W., and Herchen, H. 1963. The effect of physical form of implant on nickel sulfide tumorigenesis in the rat, *Acta Unio Int. Contra Cancrum 19*:615–619.
Gleysteen, J. J., and Stroud, R. C. 1974. Respiratory frequency tidal volume and minute volume: vertebrates, in: Altman, P. L., and Dittmer, D. S. (eds.), *Biology Data Book*, Federation of American Societies for Experimental Biology, Bethesda, Maryland, Table 208, pp. 1581–1584.
Gloyne, S. R., and Wood, W. B. 1935. Histology of tuberculous cavity wall, *Tubercle 17*:5–10.
Goldberg, A. 1966. Magnesium binding by *Escherichia coli* ribosomes, *J. Mol. Biol. 15*:663–673.
Golberg, L. 1971. Trace metal contaminants in food: Potential for harm, *Food Cosmet. Toxicol. 9*:65–80.
Goldhaber, P. 1962. Further observations concerning the carcinogenicity of subcutaneous implanted millipore filters, *Proc. Am. Assoc. Cancer Res. 4*:323.
Goldstein, A., Aronow, L., and Kalman, S. M., 1974. *Principles of Drug Action: The Basis of Pharmacology*, 2nd ed., John Wiley and Sons, New York.
Gorodiskii, V. I., Veselaya, I. V., and Rostovtseva, D. N. 1956. Cu, Zn, Cd and Ni content of muscles and tumors, Vopr. Med. Khim. 2:17–18.
Gorsuch, T. T. 1970. *The Destruction of Organic Matter*, Pergamon Press, New York.
Grabowski, C. T. 1966. Teratogenic effects of calcium salts on chick embryos, *J. Embryol. Exp. Morphol. 15*:113–120.
Greep, R. O., and Weiss, L. 1973. *Histology*, McGraw-Hill, New York.
Gross, P., and Harley, R. A. 1973. Asbestos-induced intrathoracic tissue reactions, *Arch. Pathol. 96*:245–250.
Groth, D. V. 1972. Mercury salts in chronic experiments, in: Hemphill, D. D. (ed.), *Trace Substances in Environmental Health*, Vol. 6, University of Missouri, Columbia, pp. 187–192.

REFERENCES

Gundarova, R., Lenkevich, M., and Tartakovskaya, A. 1967. *Vestn. Oftal'mol. 80*:42.
Gunn, S., Gould, A., Clark, T., and Anderson, W. A. D. 1963. Cadium-induced interstitial cell tumors in rats and mice and their prevention by zinc. *J. Nat. Cancer Inst. 31*:745–751.
Gunther, F. A., and Blinn, R. C. 1955. *Analysis of Insecticides and Acaricides,* Interscience, New York.
Gutenmann, W. H., and Lisk, D. J. 1960. Rapid determination of mercury in apples by modified Schoniger combustion, *J. Agric. Food Chem. 8*:306–308.
Gutenmann, W. H., and Lisk, D. J. 1961. Determination of selenium in oats by oxygen flask combustion, *J. Agric. Food Chem. 9*:488–489.
Guthrie, J. 1964. Histological effects of intratesticular injections of cadmium chloride in domestic fowl, *Br. J. Cancer 18*:255–260.
Haddow, A., and Horning, E. S. 1960. On the carcinogenicity of iron–dextran complex, *J. Nat. Cancer Inst. 24*:109–147.
Haddow, A., Roe, F. J. C., Dukes, C. E., and Mitchley, B. C. V. 1964a. Cadmium neoplasis: Sarcomata at the site of injection of cadmium sulfate in rats and mice, *Br. J. Cancer 18*:667–673.
Haddow, A., Roe, F. J. C., and Mitchley, B. C. V. 1964b. Induction of sarcomata in rabbits by intramuscular injection of iron dextran, *Br. Med. J. i*:1593–1594.
Halberg, I. 1969. Chronobiology, *Annu. Rev. Physiol. 31*:675–725.
Hall, T. A. 1968. Some aspects of microprobe analysis of biological specimens, in: Heinrich, K. F. J. (ed.), *Quantitative Electron Probe Microanalysis,* NBS Special Publication, 298.
Halme, E. 1961. The carcinomatous effect of zinc in drinking water, *Vitalst. Zivilisationskr. 6*:59–61.
Haro, R. T., Furst, A., Payne, W. W., and Falk, H. 1968. A new nickel carcinogen, *Proc. Am. Ass. Cancer Res. 9*:28.
Harr, J. R., Bone, J. F., Tinsley, I. J., Weswig, P. H., and Yamamoto, R. S. 1967. Selenium toxicity in rats: II. Histopathology, in: Muth, O. H. (ed.), *Symposium: Selenium in Biomedicine,* AVI Publishing Company, Westport, Connecticut, pp. 153–178.
Harr, H. R., Exon, J. H., Whanger, P. D., and Weswig, P. H. 1972. Effect of dietary selenium on N-2-fluorenyl-acetamide (FFA) induced cancer in vitamin E supplemented selenium deficient rats, *Clin. Toxicol. 5*:187–190.
Harrington, J. S. 1967. The sulfhydryl group and carcinogenesis, in: Haddow, A., and Weinhouse, S. (eds.), *Advances in Cancer Research,* Vol. X., Academic Press, New York, pp. 247–307.
Hatch, T., and Hemeon, W. C. L. 1948. Influence of particle size in dust exposure, *J. Ind. Hyg. Toxicol. 30*:175–180.
Hatch, T. F., and Gross, P. 1964. *Pulmonary Deposition and Retention of Inhaled Aerosols,* Academic Press, New York.
Hatch, W. R., and Ott, W. L. 1968. Determination of submicrogram quantities of mercury by atomic absorption spectrophotometry, *Anal. Chem. 40*:2085–2087.
Hathcock, J. N., Hill, C. H., and Matrone, G. 1964. Vanadium toxicity and distribution in chicks and rats, *J. Nutr. 82*:106–110.
Heath, J. C. 1956. The production of malignant tumors by cobalt in the rat, *Br. J. Cancer 10*:668–673.
Heath, J. C. 1960. The histogenesis of malignant tumors induced by cobalt in the rat, *Br. J. Cancer 14*:478–482.
Heath, J. C., and Daniel, M. R. 1964. The production of malignant tumors by nickel in the rat, *Br. J. Cancer 18*:261–264.
Heath, J. C., Daniel, M. R., Dingle, J. T., and Webb, M. 1962. Cadmium as a carcinogen, *Nature (London) 193*:592–593.

Heath, J. C., Freeman, M. A. R., and Swanson, S. A. V. 1971. Carcinogenic properties of wear particles from prostheses made in cobalt–chromium alloy, *Lancet 1971*:564–566.

Heath, J. C., and Webb, M. 1967. Content and intracellular distribution of inducing metal in the primary rhabdomyosarcomata induced in the rat by cobalt, nickel, and cadmium, *Br. J. Cancer 21*:768–779.

Heath, J. C., Webb, M., and Caffrey, M. 1969. Interaction of carcinogenic metals with tissues and body fluids: Cobalt and horse serum, *Br. J. Cancer 23*:153–166.

Hevesy, G., and Levi, H. 1936. Activation analysis, *Danske Vid. Selskab. Math. Fys. Medd. 14(5)*:34–40.

Hirsch, R. 1906. Über das Vorkommen von Starkekörnern im Blut, *Z. Exp. Pathol. Ther. 3*:390–392.

Hodge, H. C., and Sterner, J. H. 1943. The skin absorption of triorthocresyl phosphate as shown by radioactive phosphorus, *J. Pharmacol Exp. Ther. 79*:225–234.

Holmberg, R. E., and Ferm, V. H. 1967. Interrelationship of selenium, cadmium and arsenic in mammalian teratogenesis, *Arch. Environ. Health 18*:873–877.

Hopps, H. C. 1972. Geography, geochemistry and disease, in: Hemphill, D. D. (ed.), *Trace Substances in Environmental Health,* University of Missouri, Columbia, pp. 475–483.

Hopps, H. C., Carlisle, E. M., McKeague, J. A., Siever, R., and Van Soest, P. J. 1976. *Silicon,* National Academy of Science, Washington.

Hoste, J., Op De Beeck, J., Gijbels, R., Adams. F., Van Den Winkel, P., and de Soete, D. 1971. *Instrumental and Radio Chemical Activation Analysis,* CRC Press, Ohio.

Hueper, W. C. 1955. Cancer produced by parenterally introduced metallic Ni, *J. Nat. Cancer Inst. 16*:55–74.

Hueper, W. C. 1958. Experimental studies in metal cancerigenesis. IX. Pulmonary lesions in guinea pigs and rats exposed to prolonged inhalation of powdered metallic nickel, *Arch. Pathol. 85*:600–607.

Hueper, W. C. 1961. Carcinogens in the human environment, *Arch. Pathol. 71*:237–267;355–380.

Hueper, W. C., and Payne, W. W. 1962. Experimental studies in metal carcinogenesis: chromium, nickel, iron and arsenic, *Arch. Environ. Health 5*:445–462.

Hurley, L. S. 1968. Approaches to the study of nutrition in mammalian development, *Fed. Proc. 27*:193–198.

Hurley, L. S. and Swenerton, H. 1966. Congenital malformations resulting from zinc deficiency in rats, *Proc. Soc. Exp. Biol. Med. 123*:692–696.

Huston, J., Wallach, D. P., and Cunningham, G. J. 1952. Pulmonary reaction to barium sulfate in rats, *Arch. Pathol. 54*:430–438.

Hutcheson, D. P., Gray, D. H., Venugopal, B., and Luckey, T. D. 1975*a*. Safety of heavy metals as nutritional markers, *Environ. Qual. Saf. Suppl. 1*:74–80.

Hutcheson, D. P., Gray, D. H., Venugopal, B., and Luckey, T. D. 1975*b*. Studies of nutritional safety of some heavy metals in mice, *J. Nutr. 105*:670–675.

Hutcheson, D. P., Venugopal, B., Gray, D. H., and Luckey, T. D. 1977. Nutrient intake monitoring with lanthanide markers in humans, *Am J. Clin. Nutr.* (submitted).

I. A. E. A. 1972. *Neutron Activation Analysis Techniques in the Life Sciences,* International Atomic Energy Agency, Vienna, Austria.

Ivanov, V. I. 1965. Role of metals in deoxyribonucleic acid, *Biofizika. 10*:11.

Jackson, A. S., Michael, L. M., and Schumacher, H. S. 1972. Improved tissue solubilization for atomic absorption, *Anal. Chem. 44*:1064–1065.

Jacobs, A., and Worwood, M. 1974. *Iron in Biochemistry and Metabolism,* Academic Press, New York.

REFERENCES

Jandl, J. H., and Simmons, R. L. 1957. The agglutination and sensitization of red cells by metallic cations: interactions between multivalent metals and the red cell membrane, *Br. J. Haematol. 3*:19–22.
Jarrett, A., Spearman, R. I. C., and Riley, P. A. 1961. *Functional Dermatology,* J. B. Lippincott Co., Philadelphia.
Jasmin, G. 1963. Effects of methandrostenolone on muscle carcinogenesis induced in rats by nickel sulfide, *Br. J. Cancer 20*:190–199.
Jung, E. G., and Trachsel, B. 1970. Molekulerbiologische Untersuchungen. Zur Arsen Carcinogenese, *Arch. Klin. Exp. Dermatol. 237*:819–826.
Karasek, F. W. 1970. Ion microanalyzer, *Res. Develop. 21*:32–34.
Kasirsky, G., Gautiery, R. F., and Mann, D. E. 1965. Effect of cobaltous chloride on the minimal carcinogenic dose of methycholanthrene in albino mice, *J. Pharm. Sci 54:*491–493.
Kasprzak, K. S., and Sunderman, F. W., Jr. 1969. The metabolism of nickel carbonyl, *Toxicol. Appl. Pharmacol. 15*:295–303.
Katz, S. A. 1972. Solvent extraction for the separation of metals, *Am. Lab. 4*:19–24.
Kay, M. A., McKown, D. M., Gray, D. H., Eichor, M. E., and Vogt, J. R. 1973. Neutron activation analysis in environmental chemistry, *Am. Lab.,* July, 1973, pp. 39–48.
Kazantzis, G., and Hanburg, W. J. 1966. Induction of sarcoma in the rat by cadmium sulfide and by cadmium oxide, *Br. J. Cancer 20*:190–199.
Keberle, H. 1964. The biochemistry of desferrioxamine and its relation to iron metabolism, *Ann. N.Y. Acad. Sci. 119*:758–768.
Keller, R. 1958. Passage of bacteriophage particles through intact skin of mice, *Science 128*:718–719.
Kilburn, K. H. 1967. Cilia and mucus transport as determinants of the response of lung to air pollutants, *Arch. Environ. Health 14*:77–91.
Knorr, D. 1975. Tin—resorption, peroral toxicity and maximum admissible concentration in foods, *Lebensm. Wiss. Technol. 8*:51–56.
Kramer, H. H., and Whal, W. H. 1968. in: Wagner, H. N. (ed.), *Principles of Nuclear Medicine,* W. B. Saunders Co., Philadelphia.
Kraus, A. S., Levin, M. L., and Gerhardt, P. R. 1957. A study of occupational associations with gastric cancer, *Am. J. Public Health 47*:961–970.
Krause, W., Matheis, H., and Wolf, K. 1969. Experimentelle Fungiamie und Fungiuri durch oral Verabreichung grosser Mengen von *Candida albicans* beim gesunden Menschen (Selbstversuch), *Arzneim. Forsch. 19*:85–91.
Krehl, W. A. 1972. Mercury, the slippery metal, *Nutr. Today 7*:4–15.
Kroes, R., von Logten, M. J., and Berkrens, J. M. 1974. Study on the carcinogenicity of lead arsenate and sodium arsenate and on the possible synergistic effect of diethylnitrosamine, *Food Cosmet. Toxicol. 12*:671–679.
Krook, L. P. 1976. Inorganic macroelements, in: Recheigl, M. (ed.) *Comparative Animal Nutrition,* S. Karger, Basel, Chap. 7.
Kruger, P. 1971. *Principles of Activation Analysis,* Wiley, New York.
Kunz, J., Shahab, L., Henze, K., and David, H. 1963. The carcinogenic effect of iron–dextran, *Acta Biol. Med. Germ. 10*:602–614.
Laguillaumie, B., Champeix, J., and Jacquement, L. R. 1962. The relations between bronchial cancer and asbestosis, *Arch. Anat. Pathol. 10*:144–146.
Lamanna, C., and Sakaguchi, G. 1971. Botulinal toxins and the problem of nomenclature of simple toxins, *Bacteriol. Rev. 32*:242–249.
Landauer, W., and Sopher, D. 1970. Succinate, glycerophosphate and ascorbate as sources of cellular energy and as antiteratogens, *J. Embryol. Exp. Morphol. 24*:187–189.
Langvad, E. 1964. "Imferon"—carcinogen or co-carcinogen: Internal factors determining the

course of oncogenic virus infection, *Um. Nord. Contra Cancrum 6*:10.
Lee, A. M., and Fraumeni, J. F. 1969. Arsenic and respiratory cancer in man: An occupational study, *J. Nat. Cancer Inst. 42*:1045–1052.
Lehninger, A. L. 1975. *Biochemistry*, 2nd ed., Worth Publishers, New York.
Lenoz, W. 1962. Thalidomide and congenital abnormalities, *Lancet 2*:1332–1335.
Levina, E. N. 1966. The significance of valence for the toxicity of metals, *Zh. Ord. Vyp. Farmakol. Khimioter Sredstva. Toksikol. 54*:522.
Lindahl, T., Adams, A., and Fresco, J. R. 1966. Renaturation of transfer ribonucleic acids through site binding of magnesium, *Proc. Nat. Acad. Sci. 55*:941–948.
Lipsitiz, P. J., and English, I. C. 1967. Hypermagnesemia in the newborn infant, *Pediatrics 40*:856–862.
Litchfield, J. T., and Wilcoxon, W. F. 1949. A simplified method of evaluating dose-effect experiments, *J. Pharmacol. Exp. Ther. 96*:99–102.
Little, J. B., Kennedy, A. R., and McGandy, R. B. 1975. Lung cancer induced in hamsters by low doses of alpha radiation from polonium-210, *Science 188*:737–738.
Lockeretz, W. 1973. On the lead content of human hair, *Science 180*:1080.
Loewenstein, W. R. 1973. Membrane junctions in growth and differentiation, *Fed. Proc. 32*:60–64.
Lourenco, R. V. 1970. Distribution and clearance of aerosols, *Am. Rev. Respir. Dis. 101*:460–461.
Lourenco, R. V., Klimek, M. F., and Borowski, C. J. 1971. Deposition and clearance of 2μ particles in the tracheobronchial tree of normal subjects—smokers and non-smokers, *J. Clin Invest. 50*:1411–1420.
Luckey, M. 1974. Personal communication.
Luckey, M., Pollack, J. R., Wayne, R., Ames, B. N., and Neilands, J. B. 1972. Iron uptake in *Salmonella typhimurium* utilization of exogenous siderochromes as iron carriers, *J. Bacteriol. 111*:731–738.
Luckey, T. D. 1963. *Germfree Life and Gnotobiology*, Academic Press, New York.
Luckey, T. D. 1970. Gnotobiology is ecology, *Am J. Clin. Nutr. 23*:1533–1540.
Luckey, T. D. 1972. Introduction to intestinal microecology, *Am. J. Clin. Nutr. 25*:1292–1294.
Luckey, T. D. 1974. Introduction: The villus in chemostat man, *Am. J. Clin. Nutr. 27*:1266–1276.
Luckey, T. D. 1975. Hormology with inorganic compounds, in: Luckey, T. D., Venugopal, B., and Hutcheson, D. (eds.), *Heavy Metal Toxicity, Safety and Hormology*, Georg Thime, Stuttgart, pp. 81–103.
Luckey, T. D. 1976a. Introduction to comparative nutrition in: Recheigl, M. (ed.), *Comparative Animal Nutrition*, S. Karger, Basel.
Luckey, T. D. 1976b. Activity spectrum of ingested toxicants, in: Recheigl, M. (ed.), *Comparative Animal Nutrition*, S. Karger, Basel.
Luckey, T. D., and Stone, P. C. 1960. Hormology in nutrition, *Science 132*:1891–1893.
Luckey, T. D., and Venugopal, B. 1977. PT, a new classification system for toxic substances, *Environ. Health Perspectives* (submitted).
Luckey, T. D., Bengson, M. H., and Kaplan, H. 1974. Effect of bioisolation and the intestinal flora of mice upon evaluation of an Apollo diet, *Aerospace Med. 45*:509–518.
Luckey, T. D., Kotb, A., Vogt, J. R., and Hutcheson, D. P. 1975. Feasibility studies in rats fed heavy metals as multiple nutrient markers, *J. Nutr. 109*:660–669.
Luckey, T. D., Venugopal, B., Gray, D., and Hutcheson, D. 1977. Lanthanide marker evidence for two compartments in the human alimentary tract, *J. Nutr.* (submitted).
Lundin, P. M. 1961. The carcinogenic action of complex iron preparation, *Br. J. Cancer 15*:838–847.

REFERENCES

Lynch, K. M., McIver, F. A., and Cain, J. R. 1957. Pulmonary tumors in mice exposed to asbestos dust, *Arch. Ind. Health 15*:207–214.

Lynch, K. M., and Smith, W. A. 1935. Pulmonary asbestosis. III: Carcinoma of lung in asbesto-silicosis, *Am J. Cancer 26*:56–64.

Lyons, M., and Insko, W. M. J. 1937. Chondrodystrophy in the chick embryo produced by manganese deficiency in the diet of the hen, *K. Agric. Exp. Stn. Bull 371*:61–75.

Mailbach, H. I., Marples, R. R., and Taplin, D. 1973. Cutaneous bacteriology, in: Rook, A., (ed.), *Recent Advances in Dermatology,* Churchill Livingston, Edinburgh, Chap. 1.

Malcolm, D. 1972. Potential carcinogenic effect of cadmium in animals and man, *J. Amn. Occup. Hygiene 15*:33–36.

Manery, J. F. 1966. Effects of calcium ions on membranes, *Fed. Proc. 25*:1804.

Mao, P., and Molner, J. J. 1967. The fine structure and histochemistry of lead induced renal tumors in rats, *Am. J. Pathol. 50*:571–603.

Marcker, K., and Grade, J. 1962. A proposed structure for crystalline zinc-insulin, *Acta Chem. Scand. 16*:41.

Marples, M. J. 1965. *The Ecology of the Human Skin,* Thomas, Springfield.

Marzulli, F. N., Callahan, J. F., and Brown, D. C. 1965. Chemical structure and skin penetrating capacity of a short series of organic phosphates and phosphoric acid, *J. Invest. Dermatol. 44*:339–344.

Matson, W. R., Griffin, R. M., and Schrieber, G. B. 1970. Rapid subnanogram simultaneous analysis of zinc, cadmium, lead, copper, bismuth, and thallium, in: Hemphill, D. D. (ed.), *Trace Substances in Environmental Health,* University of Missouri Press, Columbia, pp. 396–406.

McBride, W. G. 1961. Thalidomide and congenital abnormalities, *Lancet 2*:1358.

McColl, I., and Sladen, G. E. 1975. *Intestinal Absorption in Man,* Academic Press, New York.

McCreesh, A. H. 1965. Percutaneous toxicity, *Toxicol. Appl. Pharmacol. 7*:20–26.

McLaughlin, J., Jr., Marliac, J. P., Verrett, M. J., Mutchler, M. K., and Fitzhugh, O. G. 1962. The injection of chemicals into the yolk sac of fertile eggs prior to incubation as a toxicity test, *Toxicol. Appl. Pharmacol. 5*:760–771.

Mela, L. 1968. Interactions of lanthanum and local anesthetic drugs with mitochondrial Ca^{2+} and Mn^{2+} uptake, *Arch. Biochem. Biophys. 123*:286–290.

Mello, N. K. 1975. Behavioral toxicology: A developing discipline, *Fed. Proc. 34*:1832–1834.

Mellors, R. C., and Carrol, K. G. 1961. A new method for local chemical analysis of human tissue, *Nature 192*:1090–1091.

Merck and Co. 1968. The Merck Index, 8th ed., Merck and Co., Rahway, New Jersey.

Mertz, W. 1969. Chromium in biological systems, *Physiol. Rev. 49*:163–239.

Meyers, F. H., Jawetz, E., and Goldfien, A. 1972. *Review of Medical Pharmacology,* Lang Medical Publications, Los Altos.

Miedler, L. J., and Forbes, J. D. 1968. Allergic contact dermatitis due to metallic mercury, *Arch Environ. Health 17*:960–964.

Miller, J. A. 1970. Carcinogenesis by chemicals: An overview, *Cancer Res. 30*:559–576.

Miller, L. C., and Tainter, M. L. 1944. Estimation of ED50 and its error by means of logarithmic probit paper, *Proc. Soc. Exp. Biol. Med. 57*:261–264.

Mills, C. F., and Fell, B. F. 1960. Demyelination in lambs of ewes maintained on high intakes of sulfate and molybdate, *Nature 185*:20–23.

Milner, J. E. 1969. The effects of ingested arsenic on methyl cholanthrene - induced skin tumors in mice, *Arch. Environ. Health 18*:7–11.

Mizuike, A. 1965. Separations and preconcentrations, in Morrison, G. M. (ed.), *Trace Analysis: Physical Methods,* Interscience, New York, pp. 103–159.

Moeschlin, S. 1965. in: *Poisoning: Diagnosis and Treatment,* Grune and Stratton, New York, pp. 75–93.
Morgan, J. G. 1958. Some observations on the incidence of respiratory cancer in nickel workers, *Br. J. Ind. Med. 15*:224–234.
Moschella, S. L., Pillsbury, D. M., and Hurley, H. J. 1975. *Dermatology,* W. B. Saunders Co., Philadelphia.
Moshier, R. W., and Sievers, R. E. 1965. *Gas Chromatography of Metal Chelates,* Pergamon Press, Oxford.
Muir, A. R., and Goldberg, L. 1961. The tissue response to iron–dextran: An electron microscope study, *J. Pathol. Bacteriol. 82*:471–482.
Muller, E., and Erhardt, W. 1956. Experimental contribution to the problem of the carcinogenicity of iron oxide dust, *Z Krebsforsch. 6*:65–77.
Mussman, H. C. 1975. Drug and chemical residues in domestic animals, *Fed. Proc. 34*:197–201.
N. A. S. 1974. *Chromium,* National Academy of Sciences, Washington.
Natusch, D. F. S., Wallace, J. R., and Evans C. A. 1974. Toxic trace elements: Preferential concentration in respiratory particles, *Science 183*:202–204.
Natusch, D. F. S., Wallace, J. R., and Evans, C. A. 1975. Concentration of toxic substances in submicrometer size airborne particles: The lung as a preferential absorption site, *Am. Inst. Chem. Eng. Symp. Series 147, 71*:25–32.
Neilands, J. B. 1974. *Microbial Iron Metabolism: A Comprehensive Treatise,* Academic Press, New York.
Nelson, A. A., Fitzhug, O. G., and Calvery, H. O. 1943. Liver tumors following cirrhosis by selenium in rats, *Cancer Res. 3*:230–236.
Nemethy, G., and Scheraga, H. A. 1962a. Structure of water and hydrophobic bonding in proteins. I. A model for thermodynamic properties of liquid water, *J. Chem. Phys. 36*:3382–3387.
Nemethy, G., and Scheraga, H. A. 1962b. Structure of water and hydrophobic bonding in proteins. III. The thermodynamic properties of hydrophobic bonds in proteins, *J. Phys. Chem. 66*:1773–1776.
Nothdurft, H. 1955. Experimentalle Sarkome durch reizlos einheilende Fremdkörper, *Krebsforsch. Krebsbekampf. 34*:14–27.
Nothdurft, H. 1961. Sarkomerzeugung bei Ratten durch implantierte Fremdkörper, *Reprint 8*:262-274, C. F. Boehringer and Sohne, GmbH, Mannheim.
Novey, H. S., and Martel, S. H. 1969. Asthma, arsenic and cancer, *J. Allergy 44*:315–319.
O'Dell, B. L. 1969. Effect of dietary components upon zinc availability: A review with original data, *Am. J. Clin. Nutr. 22*:1315–1322.
O'Dell, B. L. 1972. Dietary factors that affect biological availability of trace elements, *Ann. N.Y. Acad. Sci. 199*:70–81.
O'Dell, B. L., and Campbell, B. J. 1971. Trace elements: Metabolism and metabolic function, in: Florkin, M., and Stotz, E. H. (eds.), *Comprehensive Biochemistry,* Elsevier, Vol. 21, pp. 179–266.
O'Gara, R. W., and Brown, J. M. 1967. Comparison of carcinogenic action of subcutaneous implants of iron and aluminum in rodents, *J. Nat. Cancer Inst. 38*:947–952.
Oppenheimer, B. S., Oppenheimer, E. T., and Stout, A. P. 1948. Sarcomas induced in rats by implanting cellophane, *Proc. Soc. Exp. Biol. Med. 67*:33–34.
Oppenheimer, B. S., Oppenheimer, E. T., Danishefsy, I., and Stout, A. P. 1956. Carcinogenic effect of metal in rodents, *Cancer Res. 16*:439–441.
Oppenheimer, B. S., Oppenheimer, E. T., Stout, A. P., Willwhite, M., and Danishefsy, I. 1958. The latent period in carcinogenesis by plastics in rats and its relation to the precancerous stage, *Cancer (Philadelphia) 11*:204–213.

Palm, P. E., McNerneu, J. M., and Hatch, T. 1956. Respiratory dust retention in small animals. *Am. Med. Assoc. Arch. Ind. Health 13*:355–365.

Paola, J. A. di. 1964. The potentiation of lymphosarcomas in the mouse by manganese chloride, *Fed. Proc. 23*:393.

Parish, W. E., and Champion, R. H. 1973. Atopic dermatitis, in: Rook, A. (ed.), *Recent Advances in Dermatology*, C. Livingston, London, pp. 193–217.

Parsons, D. S. 1975. Energetics of intestinal transport, in: Csaky, T. Z. (ed.), *Intestinal Absorption and Malabsorption*, Raven Press, New York, pp. 9–36.

Patterson, C. C. 1965. Contaminated and natural lead environments of man, *Arch. Environ. Health 11*:344–360.

Paus, P. E. 1971. Atomic absorption, *Newsletter 10*:44.

Payne, J. M. 1964. Particulate absorption from the alimentary canal, in: Binns, T. B. (ed.), *Absorption and Distribution of Drugs*, E. and S. Livingstone, London, p. 144.

Payne, W. W. 1960. The role of roasted chromite ore in the production of cancer, *Arch. Environ. Health 1*:20–26.

Payne, W. W. 1964. Carcinogenicity of nickel compounds in experimental animals, *Proc. Am. Assoc. Cancer Res. 5*:50.

Peacock, P. R., and Peacock, A. 1965. Asbestos-induced tumors in white leghorn fowls, *Ann. N.Y. Acad. Sci. 132*:501–503.

Pearson, R. G. 1963. Hard and soft acids and bases, *J. Am. Chem. Soc. 85*:3533–3539.

Pearson, R. G. 1967. Hard and soft acids and bases, *Chem. Br. 3*:103–107.

Pearson, R. G. 1968. Hard and soft acids and bases—HSAB, Part I. Fundamental principles, *J. Chem. Ed. 45*:581–587; Part II. Underlying theories, *J. Chem. Ed. 45*:643–648.

Pham-Huu-Chanh. 1965. The comparative toxicity of sodium chromate, molybdate, tungstate and metavanadate, *Arch. Int. Pharmacodyn. Ther. 154*:243–249.

Pierre-Bienvenu, M. M., Notre, C., and Cier, A. 1963. Comparative general toxicity of metallic ions: Relation with periodic system, *C. R. Acad. Sci. 256*:1043–1044.

Pinto, S. S., and Bennet, B. M. 1963. Effect of arsenic trioxide exposure on mortality, *Arch Environ. Health 7*:583–591.

Podhrazsky, O. 1957. Lung carconoma in the steel industry, *Pracov. Lek. 9*:202–204.

Polak, L., Barnes, J. M., and Turk, J. L. 1968. The genetic control of contact sanitization to inorganic metal compounds in guinea pigs, *Immunol. 14*:707–711.

Read, W. O. 1949. Effects of germanium dioxide upon the oxygen uptake of rat tissue, Ph. D. Thesis, University of Missouri, Columbia.

Richet, C. 1906. De l'action des doses minuscules de substances sur la fermentation lactique: Périodes d'accélération et de talentissement, *Arch. Intern. Physiol. 4*:18–50.

Richmond, H. G. 1957. Induction of sarcoma in rats by an iron–dextran complex, *Scott. Med. J. 2*:169.

Richter, G. W., and Walker, G. F. 1967. Reversible association of apoferrin molecules: Comparison of light scattering and other data, *Biochem. 6*:2871–2880.

Riley, J. F. 1969. Mast cells, cocarcinogenesis and anticarcinogenesis in the skin of mice, *Experientia 24*:1237–1239.

Riviere, M. R., Chouroulenkov, I., and Guerin, M. 1960. The production of tumors by means of intratesticular injections of zinc chloride in the rat, *Bull. Ass. Fr. Etude Cancer 47*:55–87.

Robertson, D. S. E. 1970. Selenium—a possible teratogen, *Lancet I*:518–519.

Robinson, C. E. G., Bell, D. N., and Sturdy, J. H. 1960. Possible association of malignant neoplasm with iron–dextran injection: A case report, *Br. Med. J. 2*:648–650.

Robinson, J. F. 1974, Humidification, in Scurr, C., and Feldman, S (eds.), *The Scientific Foundations of Anaesthesia*, Wm. Heinemann Medical Books, London, pp. 488–496.

Robison, W. L. 1970. Electron probe microanalysis of biological materials, in: Anderson, C. A. (ed.), *Microprobe Analysis*, Wiley, New York.

Robson, A. O., and Jelliffe, A. M. 1963. Medical arsenic poisoning and lung cancer, *Br. Med. J. 2*:207–209.

Roe, F. J., and Haddow, A. 1965. Test of an iron–sorbital–citric acid complex for carcinogenicity in rats, *Br. J. Cancer 19*:855–859.

Roe, F. J., Boyland, E., and Millican, K. 1965. Effects of oral administration of two tin compounds over prolonged periods, *Food Cosmet. Toxicol. 3*:277–280.

Rohl, A. N., Langer, A. M., Selikoff, I. J., and Nicholson, W. J. 1975. Exposure to asbestos in the use of consumer spackling, patching and taping compounds, *Science 189*:551–553.

Rollinson, C. L. 1969. in: Kirschner, S. (ed.), *Coordination Chemistry*, Plenum Press, New York.

Roschin, I. V. 1967. Toxicology of vanadium compounds used in modern industry, *Gig. Sanit. 32*:26–32.

Rosenbaum, V. K. D., Wolf, E., Petermann, J., and Templin, R. 1970. Über intragastrale pH-metrie, *Zentr. Chir. 95*:153–157.

Ross, W. D., Scribner, W. G., and Sievers, R. E. 1970. in: Stock, R. (ed.), *Proceedings of the 8th International Symposium on Gas Chromatography*, Dublin.

Rostenberg, A., and Coulston, F. 1965. *Cutaneous Toxicity*, Academic Press, New York.

Roth, F. 1958. Bronchial cancer in vintners exposed to arsenic, *Virchows Arch. Pathol. Anat. Physiol. 331*:119–137.

Rothman, S. 1965. *Physiology and Biochemistry of the Skin*, The University of Chicago Press, Chicago.

Roy-Chowdhury, A. K., Mooney, T. F., and Reeves, A. L. 1973. Trace metals in asbestos carcinogenesis, *Arch. Environ. Health 26*:253–255.

Saffiotti, U., Cefis, F., and Kolb, L. H. 1968. A method for the experimental induction of bronchogenic carcinoma, *Cancer Res. 28*:104–124.

Saffiotti, U. and Wagoner, J. K. (eds.). 1976. *Occupational Carcinogenesis, Ann. N.Y. Acad. Sci. 271*:1–516.

Samitz, M. H., and Katz, S. 1964. A study of chemical reactions between chromium and the skin, *J. Invest. Derm. 43*:35–43.

Sanders, E., and Ashworth, C. T. 1961. Study of particulate intestinal absorption of hepato cellular uptake, *Exp. Cell Res. 22*:137–145.

Satterlee, H. S. 1960. The arsenic poisoning epidemic of 1900: Its relation to lung cancer in 1960: An exercise in retrospective epidemiology, *N. Engl. J. Med. 263*:676–683.

Savage, D. C., Dubos, R., and Schaedler, R. W. 1968. The gastrointestinal epithelium and its authochthonous bacterial flora, *J. Exp. Med. 127*:67–76.

Savory, J., Mushak, P., and Sunderman, F. W., Jr. 1969. Gas chromatographic determination of chromium in serum, *J. Chromatogr. Sci. 7*:674–679.

Schade, S. G., Felsher, B. F., Glader, B. E., and Conrad, M. E. 1970. Effect of cobalt upon iron absorption, *Proc. Soc. Exp. Biol. Med. 134*:741–743.

Scharpf, L. G., Hill, I. D., Wright, P. L., Plank, J. B., Keplinger, M. L., and Calandra, J. C. 1972. Effect of nitroloacetate on toxicity, teratogenicity and distribution of cadmium, *Nature 239*:231–233.

Schatz, A. 1963. The importance of metal-binding phenomena in the chemistry and microbiology of the soil. Part I. The chelating properties of lichens and lichen acids, *Advancing Frontiers of Plant Sciences 6*:113–117.

Schepers, G. W. H. 1971. Lung tumors of primates and rodents, *Ind. Med. 40*:48–53.

Schepers, G. W. H., Druham, T. M., Delehant, A. B., and Creedon, F. T. 1957. Biological action of inhaled beryllium sulfate, *Am. Med. Assoc. Arch. Ind. Health 15*:32–39.

Schinz, H. R., and Uchlinger, E. 1942. Der Metalkrebs, ein neues Prinzip der Krebserzeugung, *Z. Krebsforsch. 52*:425–437.

REFERENCES

Schinz, H. R., Baensch, W. E., Friedl, E., and Uchlinger, E. 1952. Lehrbuch der Röntgendiagnostik, *Z. Krebsforsch. 41*:628.
Schmahl, D., and Steinhoff, D. 1960. Experimental carcinogenesis in rats with colloidal silver and gold solutions, *Z Krebsforsch. 63*:586–591.
Schou, M., and Amdisen, A. 1970. Lithium in pregnancy, *Lancet 1970*:1391.
Schroeder, H. A. 1973. Recondite toxicity of trace elements, in: Hayes, W. J. (ed.), *Essays in Toxicology*, Vol. 4, Academic Press, New York.
Schroeder, H. A. 1975. *The Trace Elements and Man*, Devin-Adair Co., Greenwich.
Schroeder, H. A., Balassa, J. J., and Vinton, W. H. 1964. Chromium, lead, cadmium, nickel and titanium in mice: Effect on mortality, tumors and tissue levels, *J. Nutr. 83*:239–250.
Schroeder, H. A., and Mitchener, M. 1971. Toxic effects of trace elements on the reproduction on mice and rats, *Arch. Environ. Health 23*:102–106.
Schwarz, K. 1972. Trace elements newly identified as essential, *Proceedings of the 9th International Congress on Nutrition, Mexico*, Vol. I, Karger, Basel, pp. 96–109.
Schweitzer, G. K. 1956. The radiocolloidal properties of the rare earth elements, *U.S. Atomic Energy Commission, CRINS, 12*:31–34.
Sechzer, P. H. 1963. Cyclopropane, in: Papper, E. M., and Kitz, R. J. (eds.), *Uptake and Distribution of Anaesthetic Agents*, McGraw-Hill Book Co., New York, pp. 265–273.
Selye, H., Gabbiani, G., and Tuchweber, B. 1962. Factors influencing the development of mechanically induced experimental neoplasms, *Eur. J. Cancer 1962*:80–91.
Selye, H., and Rosch, P. J. 1954. Integration of endocrinology, in: *Glandular Physiology and Therapy*, J. B. Lippincott, Philadelphia, pp. 1–10.
Sernka, I. J. 1974. Gastrointestinal mucosa metabolism, in: Jacobson, E. D., and Shanbour, L. L. (eds.), *Gastrointestinal Physiology*, Vol. IV, University Park Press, Baltimore.
Shimoishi, Y. 1973. The determination of selenium in sea water by gas chromatography with electron capture detection, *Anal. Chim. Acta 64*:465–468.
Shin, M. L., and Firminger, H. I. 1973. Acute and chronic effects of intrapleural injection of two (2) types of asbestos in rats with a study of the histopathogenesis and ultrastructure of resulting mesotheliomas, *Am. J. Pathol. 70*;291–314.
Shubik, P., and Hartwell, J. C. 1957. Survey of compounds which have been tested for carcinogenic activity, *Suppl. 1, Public Health Publ. 149*, U.S. Government Printing Office, Washington, pp. 244–246.
Shubik, P., Saffiotie, W., Lijinsky, W., Pietra, G. W., Rappaport, H., Toth, B., Raha, C. R., Tomatis, L., Feldman, R., and Ramahi, H. 1962. Studies on the toxicity of petroleum waxes, *Toxicol. Appl. Pharmacol. 4 Suppl:* 1-62.
Sievers, R. E. 1969. in: Kirschner, S. (ed.), *Coordination Chemistry*, Plenum Press, New York.
Singer, B., and Frankel-Conrat, H. 1962. Enzyme resistance of complexes of ribonuclear acids with metals, *Biochemistry 1*:852–858.
Skoryna, S. C., and Waldron-Edward, D. (eds.) . 1971. *Intestinal Absorption of Metal Ions, Trace Elements and Radionuclides*, Pergamon Press, New York.
Smith, S. D. 1974. Effects of electrode placement on stimulation of frog limb regeneration, *Ann. N.Y. Acad. Sci. 238*:500–507.
Sollman, T. 1948. *A Manual of Pharmacology*, 7th ed., W. B. Saunders Co., Philadelphia.
Somers, E. 1960. Fungitoxicity of metal ions, *Nature 187*:427–428.
Spencer, H., Greenberg, J., Berger, E., Perrone, M., and Lazlo, D. 1956. Studies of the effect of ethylenediamine tetraacetic acid in hypercalcemia, *J. Lab. Clin. Med. 47*:29–41.
Spyker, J. M. 1975. Assessing the impact of low-level chemicals on development: Behavioral and latent effects, *Fed. Proc. 34*:1835–1844.
Stanton, M. F. 1967. Primary tumors of bone and lungs in rats following local deposition of

copper chelated *N*-hydroxy-2-acetylaminofluorene, *Cancer Res.* 28:1000–1006.

Steffee, H., and Baetjer, A. M. 1965. Histopathologic effects of chromate chemicals: Report of studies in rabbits, guinea pigs, rats and mice, *Arch. Environ. Health 11*:66–75.

Stockinger, H. E. 1966. *Beryllium, Its Industrial Hygiene Aspects*, Academic Press, New York.

Stockinger, H. E., and Coffin, D. L. 1968. Biological effects of air pollutants, in: *Air Pollution*, Academic Press, New York.

Stocks, P. 1960. On the relation between atmospheric pollution in urban and rural localities and mortality from cancer, bronchitis and pneumonia with particular reference to 3,4-benzopyrene, beryllium, molybdenum, vanadium and arsenic, *Br. J. Cancer 14*:397–410.

Stocks, P., and Davies, R. I. 1960. Epidemiological evidence from chemical and spectrographic analysis that soil is concerned in the causation of cancer, *Br. J. Cancer 14*:8–22.

Stone, O. J. 1969. The effect of arsenic on inflammation, infection and carcinogenesis, *Tex. State J. Med. 65(10)*:40–43.

Sunderman, F. W., and Donnelly, A. J. 1965. Studies of nickel carcinogenesis: Metastasizing pulmonary tumors in rats induced by the inhalation of nickel carbonyl, *Am. J. Pathol. 46*:1027–1041.

Sunderman, F. W., and Sunderman, F. W., Jr. 1961. Nickel poisoning: Implication of nickel as a pulmonary carcinogen in tobacco smoke, *Am. J. Clin. Pathol. 35*:203–209.

Sunderman, F. W., Jr. 1967. Inhibition of induction of benzoprene hydroxylase by nickel carbonyl., *Cancer Res. 27*:950–955.

Sunderman, F. W., Jr. 1968. Nickel carcinogenesis, *Dis. Chest 54*:527–534.

Sunderman, F. W., Jr. 1971. Metal carcinogenesis in experimental animals, *Food Cosmet. Toxicol. 9*:105–120.

Suzuki, K. 1975. Study on dermal excretion of metallic elements (Na, K, Ca, Mg, Fe, Mn, Zn, Cu, Cd, Pb), *Xth Internat. Cong. Nutr. 1975*:1520 (abstract).

Swenerton, H., and Hurley, L. S. 1974. Teratogenic effects of a chelating agent and their prevention by Zn, *Science 173*:62–63.

Swenerton, H., Shrader, L., and Hurley, L. S. 1969. Zinc deficient embryos: reduced thymidine incorporation, *Science 166*:1014–1015.

Szabo, K. T., Hawk, A. M., and Henrey, M. 1970. The teratogenic effect of lithium carbonate upon the palate of random bred mice, *Toxicol. App. Pharmacol. 17*:274.

Szent-Gyorgi, A., Isenberg, I., and Baird, S. L., Jr. 1960. Electron donating properties of carcinogens, *Proc. Nat. Acad. Sci. 46*:1444–1449.

Talbot, R. B., Davison, F. G., Green, J. W., Reece, W. O., and Van Gelder, G. 1965. Effects of subcutaneous implantation of rare earth metals, *U.S.A. E.C. Report 1170*.

Tanami, J. 1960. Infection and germfree animals, *Mod. Med. 6*:287–299.

Tapp, E. 1966. Changes in rabbit tibia due to direct implantation of beryllium salts, *Arch. Pathol. 88(5)*:521–529.

Taylor, M. L., Arnold, E. L., and Sievers, R. E. 1968. Rapid microanalysis of beryllium in biological fluids by gas chromatography, *Anal. Lett. 1*:735.

Thomas, J. A., and Thiery, J. P. 1953. Production élective de liposarcoma chez des lapins par les oligo éléments zinc et cobalt, *C. R. Acad. Sci. 236*:1387–1389.

Tolug, G. 1972. Extreme trace analysis of the elements. 1. Methods and problems of sample treatment, separation and enrichment, *Talanta 19*:1489.

Toner, R. G., Carr, K. E., and Wyburn, G. M. 1971. *The Digestive System—An Ultrastructural Atlas and Review*, Appleton–Century–Crofts, New York.

Toremalm, N. G. 1960. The daily amount of tracheo-bronchial secretions in man, *Acta Otolaryngol. Suppl. 158*:43–53.

REFERENCES

Trapmann, H., and Devani, M. 1965. Der katalytische Effekt von Thorium- und Zirkonium-Ionen auf die Hydrolyse, *Z. Physiol. Chem. 340*:81–85.

Treytl, W. J., Orenberg, J. B., and Marich, K. W. 1972. Detection limits in analysis of metals in biological materials by laser microprobe optical emission spectroscopy, *Anal. Chem.* 44:1903–1904.

Trevan, J. W. 1927. The error of determination of toxicity, *Proc. R. Soc. B. 101*:483–484.

Tscherkes, L. A., Aptekar, S. G., and Volgarev, M. N. 1961. Hepatic tumors induced by selenium, *Byull. Eksp. Biol. Med. 53*:78–82 (Russian).

Tscherkes, L. A., Volgarev, M. N., and Aptekar, S. G. 1963. Selenium-caused tumors, *Acta Unio Int. Contra. Cancrum 19*:632–633.

Tuchmann-Duplessis, H., and Mercier-Parot, L. 1956. Influence d'un corpuscle chelation, l'acid ethylenediaminetetraacetique sur la gestation et le developpement foetal du rat, *C. R. Acad. Sci. 243*:1064–1066.

Tupper, R., Watts, R. W. E., and Normall, A. 1955. The incorporation of Zn in mammary tumors and some other tissues of mice after injection of the isotope, *Biochem. J. 59*:264–268.

Turner, F. C. 1951. Sarcomas at sites of subcutaneously implanted bakelite disks in rats, *J. Nat. Cancer Inst. 2*:81–83.

Ulrich, J. A. 1965. Skin carriage of bacteria in the human, in: *Spacecraft Sterilization Technology* (NASA SP-108), NASA, Washington D.C., pp. 87–96.

Underwood, E. I. 1971. *Trace Elements in Human and Animal Nutrition*, 3rd ed., Academic Press, New York.

Utidjian, H. M. 1973. Criteria for a recommended standard. Occupational exposure to beryllium and its compounds, *J. Occup. Med. 15*:659–662.

Van Esch, G. J., and Kroes, R. 1969. The induction of renal tumors by feeding basic lead acetate to mice and hamsters, *Br. J. Cancer 23*:765–771.

Van Esch, G. J., van Genderen, E., and Vink, H. H. 1962. The induction of renal tumors by feeding basic lead acetate to rats, *Br. J. Cancer 16*:289–297.

Van Peenen, H. J. 1966. *Essentials of Pathology*, Year Book Medical Publishers, Chicago.

Venugopal, B., and Luckey, T. D. 1977. *Metal Toxicity in Mammals*. Vol. 2. *Toxicity of Metals and Their Compounds in Mammals.*, Plenum Press, New York.

Verzar, F. 1967. *Absorption from the Intestine*, Hafner Publishing Co., New York.

Volkheimer, G. 1974a. *Persorption*, George Thieme Verlag, Stuttgart.

Volkheimer, G. 1974b. Passage of particles through the wall of the gastrointestinal tract, *Environ. Health Prospect. 9*:215–225.

Vorwald, A. J., Reeves, A. L., and Urban, E. C. J. 1966. Experimental beryllium toxicology, in: Stockinger, H. E. (ed.), *Beryllium: Its Industrial Hygiene Aspects*, Academic Press, New York.

Wacker, W. E. C., and Vallee, B. L. 1959. Nucleic acids and metals. I. Cr, Mn, Ni, Fe and other metals in ribonucleic acid from diverse biological sources, *J. Biol. Chem. 234*:3257–3262.

Walsh, A. 1955. Application of atomic absorption spectra to chemical analysis, *Spectrochim. Acta 7*:108–117.

Walters, M., and Roe, F. J. C. 1965. A study of the effects of tin and zinc admintered orally to mice over a prolonged period, *Food Cosmet. Toxicol. 3*:271–276.

Warden, A. N., and Harper, K. H. 1964. Oral toxicity as influenced by method of administration, *Proc. Exp. Soc. Study Drug Toxicity 4*:107–110.

Waters, M. D., Gardner, D. E., Aranyi, C., and Coffin, D. L. 1975. Metal toxicity for rabbit alveolar macrophages *in vitro, Environ. Res. 9*:32–47.

Waxman, H. S., and Brown, E. B. 1969. in: Brown, E. B., and Moore, C. V. (eds.), *Progress in Hematology*, Vol. 6, Grune and Stratton, New York.
Weaver, J. C., Koslainsek, V. M., and Richards, P. D. N. 1956. Cobalt tumor of thyroid gland, *Calif. Med. 85*:110–112.
Webb, M., Heath, J. C., and Hopkins, T. 1972. Intranuclear distribution of the inducing metal in primary rhabdomyosarcomata induced in the rat by Ni, Co and Cd, *Br. J. Cancer 26*:274–278.
Wedderburn, J. F. 1972. Selenium and cancer, *N. Z. Vet. J. 20*:56.
Wegener, W. S., and Romano, A. H. 1963. Zinc stimulation of RNA and protein synthesis in *Rhizopus nigricans*, *Science 142*:1669–1670.
Weil, C. S. 1952. Tables for convenient calculation of median-effective dose (LD_{50} or ED_{50}) and instructions in their use, *Biometrics 8*:249–263.
Weil, C. S., and Rostenberg, A. 1969. *Evaluation of Safety of Cosmetics*, Academic Press, New York.
Weinberg, E. D. 1974. Iron and susceptibility to infectious disease, *Science 184*:952–956.
White, A., Handler, P., and Smith, E. L. 1973. *Principles of Biochemistry*, 5th ed., McGraw-Hill Book Co., New York.
Whitmore, C. E., and Huebner, R. J. 1972. Inhibition of chemical carcinogenesis by viral vaccines, *Science 177*:60–61.
Wilkinson, P. S. 1965. Contact dermatitis in man with special reference to ecologic factors, in: Brook, A. J., and Walton, G. S. (eds.), *Comparative Physiology and Pathology of the Skin*, F. A. Davis Co., Philadelphia, pp. 521–536.
Williams, R. B. 1976. Trace elements, in: Recheigl, M. (d.), *Comparative Animal Nutrition*, S. Karger, Basel, Chap. 8.
Williams, R. M., and Beck, F. 1969. A histochemical study of gut maturation, *J. Anat. 105*:487–501.
Wilson, I. B., and La Mer, V. K. 1948. The retention of aerosol particles in the human respiratory tract as a function of particle radius, *J. Ind. Hyg. Toxicol. 30*:265–280.
Witschi, H. P., and Aldridge, W. N. 1968. Uptake, distribution and binding of beryllium to organelles of the rat liver, *Biochem. J., 106*:811–817.
Woldseth, R. 1973. *X-Ray Energy Spectrometry*, Kevex Corp., Burlingame.
Wood, J. M. 1974. Biological cycles for toxic elements in the environment, *Science 183*:1049–1052.
Zeedijk, N. 1952. Experimentelle Erzeugung maligner Nieren Kapseltumoren bei der Ratte durch Druckreiz, *Schweiz. Z. Pathol. Bakteriol. 15*:666–671.
Zeedijk, N. 1973. Investigation of asbestos bodies and asbestos fibers found in the lungs of a mesothelioma patient by electron microscopy, *Mikrochim. Acta 6*:977–984.
Zollinger, H. W. 1953. Kidney adenomas and carcinomas in rats caused by chronic lead poisoning and their relationship to corresponding human neoplasma, *Virchow's Arch. Pathol. Anat. 323*:694–710.

INDEX

Absorption, 15, 19, 39, 43-91, 94, 95, 98-101, 108, 110, 113, 117, 132, 161, 164, 169-191
Acid, 112
Actinide, 111, 118, 133, 137, 172, 175, 191
Activation, 121, 122
Active transport, 42, 53, 54, 84, 108
Activity spectrum, see Biologic activity spectrum
Aerodynamics, 59-74
Aerodynamic diameter, 65-72
Aerosol, 64, 67, 68, 149
Affinity, 111, 112
Agglutination, 108-111, 120-122, 128
Air velocity, 59, 60, 68-72
Alkali disease, 2
Alkaline earth metals, see Metals, alkaline earth
Allosteric, 122
Aluminum (Al), 29, 35, 96, 107, 123, 137, 156, 159, 171, 172, 190
Alveolar volume, 61, 71, 76-80
AMP, 124, 125
Anemia, 166, 169, 172, 177, 178, 183, 184
Anion effect, 107
Antagonism, 96, 146
Anorexia, 172, 173, 177, 180, 183
Anticoagulant, 174
Antimetabolite, 178
Antimony (Sb), 120, 139, 179
Antiseptic, 12
Apoferritin, 100, 122
Argyria, 95, 125, 167
Arsenic (As), 2, 6, 13, 21, 35, 82, 84, 96, 97, 120, 131-133, 139, 140, 160, 161, 178, 179, 182

Arthritis, 12
Asbestos, 107, 132-139, 149-151
Astringent, 12
Ataxia, 170, 185
ATP, 42, 124, 125
ATPase, 170, 180, 184

Bacteria, see Microbes
Barium (Ba), 11, 12, 21, 35, 91, 167-169,
Base, 112
Basic, basicity, 109
Beryllium (Be), 21, 24, 32, 84, 85, 98, 116, 120, 122, 124, 132, 135, 146, 147, 149, 156, 158, 163, 167-169, 191
Bidentate, 115
Binding, irreversible, 126
Biologic activity spectrum, 3, 6, 14, 15
Biosphere, 2
Biotransformation, 103, 130, 142, 145, 148
Biphasic mortality, 174
Biphasic response, 14
Bismuth (Bi), 13, 35, 95, 109, 122, 123, 139, 180, 190, 192
Blood dyscrasia, 167
Blood–gas exchange, 62
Bone, 144, 160, 172, 175, 188
Brain, 166, 169, 171, 176, 177, 182, 184, 187
Brush border, see Microvilli

Cadmium (Cd), 5, 6, 24, 28, 33, 35, 84, 93, 96, 98, 107, 111, 121, 123, 124, 132, 136, 138, 146, 147, 158, 160, 161, 169, 170, 182
Caffein, 57
Calcification, 168, 173, 174, 182

233

Calcium (Ca), 12, 29, 41, 42, 54, 102, 111, 112, 120, 122, 123, 127, 138, 158, 168, 177, 178, 183, 191
cAMP, 165
Cancer, 129-153
  also see Tumor
Cardiac failure, 168, 173, 174, 179, 180, 183, 185
Carrier protein, 42, 54, 148
Cell degeneration, 166
Central nervous system, *see* Brain
Cerium (Ce), 126, 133, 137, 174
Ceruloplasmin, 117, 122, 183
Cesium (Cs), 164, 166, 190, 192
Charge transfer complex, 146
Chelate, chelation, 4, 28, 33. 51, 52, 56, 95, 97-100, 104, 115-120, 123-125, 155, 164, 169, 185, 187
Chemotherapeutic index, 12
Chlorophyll, 118
Chromium (Cr), 2, 13, 18, 23, 26, 29, 32, 42, 84, 96-98, 102, 107, 116, 122, 132, 133, 137, 138, 141, 142, 147, 148, 160, 163, 180, 181, 183
Chronotoxicology, 41
Chrysolite, 138
Chrysotile, *see* Chrysolite
Chyme, 47, 49, 51, 55, 56
Cilia, 62, 63, 69, 73, 78
Circulatory failure, 165, 168, 173, 185
Clearance, 88
Cobalt (Co), 2, 11, 13, 31, 32, 42, 54, 84, 111, 120, 121, 123, 124, 132, 133, 143, 144, 146, 148, 149, 158, 160, 182, 185, 186
Cocarcinogen, 132, 134, 136, 143, 178, 186
Colligative properties, 41, 53, 107, 110, 163, 167
Colloid, 98, 110, 149, 164, 180, 181
Colloidoclastic shock, 87, 172, 176, 180
Conalbumin, 122
Conjunctivitis, 187
Convulsions, 165, 166, 173, 177, 180, 182, 185
Coordination, 115, 116
Copper (Cu), 2, 12, 29, 33, 34, 84, 89, 96, 98, 104, 111, 117, 120-125, 132, 133, 135, 156, 158, 164, 166, 169, 170, 177, 183
Coprecipitation, 108
Choreal movement, 187

Covalent, 111, 115
Crocidolite, 138
Cyanosis, 168

Death, 2, 6, 7, 14, 15, 52, 103, 154, 159, 163-173, 180-187
Delayed toxicity, *see* Latent period
Demyelination, 160
Denaturation, 120-122, 163
Dental caries, 182
Deoxynucleic acid, DNA, 15, 123-126, 140, 145, 147, 163, 168
Deposition, particulate, 70-74
Dermatitis, 86, 122, 140, 167, 172, 183, 187, 188
Desferrioxamine, 120
Desmosome, 47, 57
Desquamation, 101
Detoxication, 93-97, 104, 117, 119, 155, 161, 164, 165, 170, 183, 186
Diarrhea, 165-172, 179-186
Diet, 2, 4
Differentiation, 125, 135, 150, 155, 158, 159, 163, 168
Diffusion, 41, 54, 84, 85, 108
Digesta, *see* Chyme
Disease, 4
Diversity, 3
DNA, *see* Deoxynucleic acid
Dose–response, 3, 6,9, 22, 25, 39, 80
Dynamic equilibrium, 41, 104, 147, 148, 163, 189, 191, 192
Dysprosium (Dy), 35, 137, 174

Ecology, 15, 163
Edema, 179, 186
EDTA, 95, 97, 98, 115, 118, 119, 174
Electrochemical, 15, 106
Electronegativity, 112, 115, 131, 150, 169
Electron mobility, 112, 146.
Electron transport, 106, 117, 167, 171
Electrophilic, 146, 152
Electrophoretic mobility, 127
Electropositive, 106, 112, 150, 164, 167, 169, 172, 174, 176, 181, 184, 185, 188-192
Electrostatic, 115
Electrovalent, 111, 188
Embryocidal, 155-160
Embryogenesis, 154
Emphysema, 169, 174

INDEX

Endocytosis, 41-43, 55, 56, 147
Enterohepatic, 40, 72, 170
Environment, 1, 2, 3, 6, 14, 15, 130, 149, 150, 153, 161, 163, 170, 171
Enzyme inhibition, 163, 179, 180, 186
Epithelial cells, 63
Essential metal, see *Metal,* essential
Essential nutrient, *see* Nutrient, essential
Europium (Eu), 35, 175
Evolution, 3
Excitation, *see* Stimulation
Excretion, 40, 64, 95-104, 117, 119, 161, 164, 173-176, 180-187
Exhaustion, 170, 179
Extrusion zone, 48, 58

Facilitated diffusion, *see* Facilitated transport
Facilitated transport, 42, 53, 54, 108
Ferrioxamine, 118
Ferritin, 100, 115, 122
Ferrocene, 123
Fertilization, 153
Fever, 181
Fibrosis, 74, 151, 152, 169
Free radical, 152, 153
Fume fever, 169

Gadolinium (Gd), 31, 109, 133, 137, 174
Gallium (Ga), 97, 137, 171, 172
Gastroenteritis, 173, 180-186
Geology, 2
Germanium (Ge), 13, 22, 31, 35, 96, 137, 159, 176
Germfree, 51, 64
Glutathione, 96, 128, 166
Glycocalyx, 44-49, 54-57
GMP, 125
Gold (Au), 12, 35, 84, 120, 122, 125, 132, 135, 149, 164, 166, 167, 190
Granuloma, 73, 151-153, 168, 172-176
Growth, 6, 15, 24, 148, 154, 163, 169, 172-187
Growth promotant, 9

Hafnium (Hf), 109, 137, 178, 190
Half-life, 76
Hallucination, 170
Hapten, 122, 152
Hard acid, *see* HSAB
Hard base, *see* HSAB

Health, 1, 20-26
Hemachromatosis, 185
Heme proteins, 118
Hemorrhage, 174, 176, 186
Hemolysis, 179, 187
Hemosiderosis, 185
Hepatic, *see* Liver
Histogenesis, 153
Homeostasis, 14, 41, 61, 64, 73, 81, 84, 93, 99-103, 122, 154, 155, 161-169, 177, 184, 185
Hormetic, 9, 22, 23, 25
Hormology, 8-11, 23
Hormone, 101, 117
Hot spots, 175
HSAB (Hard and soft acid and base theory), 3, 111-114
Hydration, 107-109
Hydrocephalus, 158, 160, 183
Hydrolysis, 95, 97, 108-111
Hydroxylation, 97
Hyperemia, 179, 181
Hypermagnesemia, 158
Hypertension, 165, 166, 169, 173

Immune system, 122, 151, 177, 178, 180
Implantation, 153
Implosion, 45, 58
Indium (In), 5, 35, 96, 131, 133, 159, 161, 171-173, 192
Infection, 132
Inhalation, 19, 40, 59-80
Ion, 110-114, 164
Ion-dipole, 108
Ionization, 108, 109, 155
Iridium (Ir), 185, 187, 190
Iron (Fe), 2, 13, 29, 35, 41, 54, 100-104, 107, 115-117, 120-124, 132, 135, 138, 143, 144, 147, 160, 166, 167, 170, 177, 178, 184-186
Iron dextrin, 132, 143, 144, 152, 186
Isotope, 98
Itai-itai-byo, 170

Kidney, 139, 140, 163-187

Lanthanide, 23, 95, 96, 110, 111, 113, 118, 120, 123, 126, 171, 174, 175, 190
Lanthanum (La), 50, 174, 190
Latent period, 130, 138, 141-146, 148-152, 163, 168

Lead (Pb), 5, 13, 26, 33-35, 72, 95-98, 108, 111, 121, 123, 132, 133, 139, 159-161, 170, 173, 176-179, 188-191
Lifeterm studies, 7, 16, 23, 87
Ligand, 4, 51, 104, 111-120, 123, 127, 131, 164, 173, 178
Limb bud, 159, 173
Lithium (Li), 7, 11, 12, 33, 35, 108, 116, 158, 164, 165, 189, 1-91
Liver, 142, 158, 166-169, 172-179, 182-187
Longevity, 142
Lung, 39, 40, 59-80, 95, 135, 137-145, 151, 152, 163, 167, 169-176, 180, 181, 184-187
Lutetium (Lu), 133
Lysis, 128
Lysosome, 96, 119, 122, 125, 128, 147, 148, 163, 166

Macrophage, 39-41, 56, 57, 72, 73, 85, 96, 137, 152, 186
Magnesium (Mg), 12, 28, 29, 67, 101, 102, 107, 111, 112, 115, 120, 123-127, 132, 135, 138, 147, 158, 167-169, 189-191
Malformation, 154
Mammary, 136
Manganese (Mn), 13, 29, 35, 54, 75, 107, 111, 112, 115, 120, 123-125, 132, 143, 155, 156, 180, 184, 185
Manganism, 184
Maturation, functional, 153, 154, 163
Megalopolophilic, 2
Membranes, 111, 126-128, 154, 163, 165, 171, 175, 176, 178, 183
Mercury, (Hg), 5, 12, 15, 16, 21, 24, 26, 28, 34, 81-84, 93-98, 102, 107, 108, 111, 120-125, 127, 131, 132, 136, 156-162, 167, 169-173, 182, 188-191
Metabolism, 173
Metal, cofactor, 121
  essential, 5-9, 25, 117, 145-148, 154, 156, 163, 180, 191
  trace, 1, 116, 138, 145, 147, 150, 154, 180
  transitional, 105-107, 115, 118, 124
Metallic taste, 173
Metallocene, 146
Metalloenzyme, 115
Metallothionein, 93, 96, 115, 170
Metal prosthesis, 149

Metals, several or all, 8, 36, 64, 66, 99, 105, 106, 113, 114, 119, 128, 131, 134, 157
  alkaline earth, 105-107, 118, 123, 124, 127
  transuranium, 175
Methylcholanthrene, 144
Methyl mercury, 125, 158, 159, 162
Microbe, microorganism, 3, 39-43, 46-53, 57, 61, 64, 67, 83, 162, 170
Microcalyx, *see* Glycocalyx
Microcolloidal, *see* Radiocolloidal
Microvilli, 42-48, 54, 55, 100
Minamata, 159
Minute volume, 59-62, 78-80
Mitosis, 154
Mitotic poison, 173
Molybdenum (Mo), 18, 33, 141, 160, 166, 181, 183, 184, 191
Monocytes, *see* Macrophage
Mortality, *see* Death
Mucosa, 44, 45, 54-56, 95, 100
Mucus, 68, 69
Multidentate, 115
Muscle, 59, 82, 88, 89, 144, 166, 169, 173
Muscle paralysis, 169, 183
Mutation, 154

Nature, 6
Nephrotoxic, 98
Nerve, 144, 165, 166, 168, 171-173, 180, 183
Nickel (Ni), 2, 7, 29, 33, 34, 54, 81, 84, 111, 112, 116, 120-126, 132, 133, 137, 138, 143-149, 160, 185, 186
Nickel carbonyl, 145, 187
Nickelocene, 145, 146
Niobium (Nb), 109, 132, 139, 178, 181
Nuclei, anaplastic, 152
Nucleic acid, 103, 123-126, 145, 146, 163
Nucleoli, 146
Nucleophilic, 146
Nurture, 5, 6
Nutrient, essential, 1, 2, 6-8, 11, 25, 99, 132, 137, 154
Nutrition, 3, 5-8, 11, 14

Occupational hazard, 1, 4, 14, 24, 37, 77, 132, 135, 138, 140, 141, 144, 161, 170
Olation, 97, 104, 108-111, 176, 180, 181
Oligodynamic action, 9

INDEX

Oncogenic, *see* Tumor and Surface oncogenesis
Ontogeny, 151
Organogenesis, 153, 154
Organometallic compounds, 5, 13, 42
Osmium (Os), 35, 95, 98, 125, 185, 187, 190
Osteomalacia, 170
Oxolation, 110
Oxyacid, 180, 184, 189, 192
Oxysalt, 56, 95, 178, 180, 190

Palladium (Pd), 132, 143, 145, 185, 187
Paralysis, 169, 176, 186, 187
Particle, 23, 43, 54-57, 59-62, 66-75, 78, 86, 98, 104, 107-111, 150, 151
Passive transport, 41, 53-55
Penicillin and derivatives, 118, 120
Peptizing agent, 110
Perimatrix, *see* Glycocalyx
Peritonitis, 175
Permeability, 41, 103, 127, 128, 178
Persorption, 41, 43, 46, 57, 58
pH, 43, 54, 56, 85, 108, 110, 117, 125, 171, 174, 176, 181, 184
Phagocytosis, 41, 46, 57, 58, 62, 72, 78, 85-89, 95, 109, 111, 151, 152, 175-178, 186
Piezoelectric, 151, 153
Pinocytosis, 41, 46, 49, 54-56
Placental membrane, 154-160, 177, 182, 183
Platinosis, 188
Platinum (Pt), 11, 116, 133, 185, 188, 190
Plutonium (Pu), 21, 120, 133
Pneumonitis, 169, 179, 181
Polarity, cell, 151
Pollution, 4, 23, 64-69, 75, 81, 150, 161
Polonium (Po), 132, 133
Polycythemia, 186
Polyelectrolyte, 123
Polymerization, 109, 111, 117, 121, 123, 176
Potassium (K), 29, 41, 42, 98, 102, 133, 159, 165, 166, 168, 171, 173, 190, 191
Praseodymium (Pr), 28, 33, 137, 174
Premalignant cells, 152, 153
Prostigmin, 57
Prostration, 179, 185
Protein, 103, 120-122, 126-128, 146-148, 178, 182

Psychosis, maniac, 158
pT, 19, 20, 21

Quartz, *see* Silica

Radiation, 132, 133, 137, 154, 175, 192
Radiocolloid, 98, 104, 110, 111, 164
Radium, 169
Recondite toxicity, 132, 163
Reproduction, 15, 24, 163, 165, 169, 177, 178, 182, 183, 186
Resorption, 154, 158, 159, 165, 171, 179
Respiratory disease, 77
Respiratory failure, 167, 168, 173, 174, 180, 182, 184
Respiratory irritant, 64-76
Retention, particulate, 70-76
Reticuloendothelial system, 95-97, 111, 148, 171-175
Rhenium (Re), 184, 185, 187, 190
Rhodium (Rh), 132, 143, 145
Ribonucleic acid, RNA, 123-126, 140, 144, 145, 146, 147, 148
Ribosomes, 126
Rickets, 184
RNAse, 125
RNA polymerase, 135, 147
Rubidium (Rb), 133, 164, 165, 166, 185, 187, 190

Safety, 1, 2, 4, 5, 20-26
Samarium (Sm), 133
Scandium (Sc), 132, 137, 173
Sequester, 95, 100-104
Selenium (Se), 2, 6, 13, 16, 23, 26, 32-35, 64, 72, 96, 132, 141-143, 147, 160, 161, 167, 171, 181, 190
Selenomethionine, 142, 147
Sensitivity, 27-37, 41, 86, 122
Settling velocity, 68
Siderochrome, 51
Silica and silicates, 74, 137-139, 149-151
Silicon (Si), 6, 7, 29, 31, 35, 74, 85, 95, 107, 132, 137-139, 149, 150
Silver (Ag), 11, 12, 29, 33, 84, 89, 95, 96, 124, 125, 132, 135, 149, 164, 166, 167
Skin, 13, 39, 80-86, 101, 140, 170, 172, 179-183, 186, 187
Sodium (Na), 9, 21, 29, 35, 41, 42, 98, 102, 138, 165, 190, 191

Soft acid, *see* HSAB
Soft base, *see* HSAB
Solubility, 107
Sterile, 158, 169
Stimulation, 2, 3, 6, 8-15, 22, 39, 46, 57, 104, 142, 153, 168, 170, 174, 179, 191, 192
Stress, 4, 26, 41, 148, 151, 163
Strontium (Sr), 7, 29, 54, 124, 168, 169
Succus entericus, 45, 46, 49
Surface charge, 127
Surface oncogenesis, 129, 132, 136-139, 145, 148-153
Synergism, 130, 132

Taconite, 143
Tantalum (Ta), 74, 132, 139, 149, 181, 190
Technetium (Tc), 13, 184, 185
Tellurium (Te), 97, 132, 143, 160, 181, 182, 190
Teratogenic, 120, 153-165, 169, 171, 173, 175, 179, 182
Terminal web, 47, 55, 57
Testes, 136, 169, 177, 183
Tetracycline, 118, 120
Tetraethyl lead, 139
Thalidomide, 153
Thallium (Tl), 5, 11, 21, 35, 74, 81, 84, 96, 108, 131, 159, 161, 171-173, 182, 188-192
Therapy, 4, 6, 10-15, 86, 119, 140, 142
Thorium (Th), 133, 175
Thymus, 147
Thyroid, 144
Tin (Sn), 2, 6, 7, 84, 132, 139, 149, 159, 176, 177
Titanium (Ti), 13, 32, 33, 132, 159, 176, 178
Titanocence, 132, 146

TMP, 125
Tolerance, 16
Tonsils, 57
Trace metal, *see* Metal, trace
Tracer, radioactive, 98
Transferrin, 96, 101, 122
Transitional metals, *see* Metal, transitional
Transuranium metals, *see* Metals, transuranium
Tubercle, 74, 78
Tumor, 104, 129-153, 168, 169, 172-174, 177-180, 182, 185-187, 191
Tungsten (W), 18, 33, 141, 181, 183, 184, 190

Ulceration, 166, 170, 179, 183
Unity, 3
Unwinding, polymeric, 123, 124
Uranium (U), 75, 122, 124, 125, 127, 175
Uranyl, *see* Uranium

Vanadium (V), 2, 7, 11, 13, 29, 33, 96, 97, 104, 116, 133, 139, 140, 160, 178, 180
Variation, individual, 3
Villus, 44-50, 55, 57
Vomit, 43, 166, 172, 179, 181

Water, monomeric, 108
Wilson's disease, 166

Ytterbium (Yb), 137, 174
Yttrium (Y), 132, 136, 173, 174

ZEP (Zero equivalent point), 10, 22-24
Zero tolerance, 22
Zinc (Zn), 2, 12, 29, 33, 35, 67, 84, 96, 99, 107, 111, 115, 116, 120-126, 132, 135, 136, 148, 154, 158, 166, 169, 170, 180, 183, 189
Zirconium (Zr), 13, 85, 109, 132, 138, 178